Cracking the Code: International Strategies for Preventing and Managing Alkali-Aggregate Reaction

Shalini

Table of Contents

1. Introduction

1.1 Problem Statement

Alkali Aggregate Reaction (AAR) is a deterioration mechanism which affects concrete structures all over the world. Different parts of the world employ various mitigation and control measures for AAR damage. Different tests are also performed worldwide to assess AAR. With the variety of AAR avoidance measures and AAR tests performed worldwide, it is necessary to have a thorough compilation and critical assessment of these AAR avoidance measures and AAR tests, which may be of assistance to engineers and other professionals who are involved in structural and material design of concrete structures or in the construction, quality control and condition monitoring and assessment of concrete structures.

1.2   Background

AAR in concrete refers to a group of chemical reactions which take place between the alkalis within the Portland cement paste and certain reactive forms of silica containing minerals in the aggregates (Godart & de Rooij, 2013). The alkalis in the concrete pore solution come from the hydration process of manufacturing clinker in the production of cement, although they may be supplemented by alkalis that are found in some aggregates or alkaline salts from the environment. In the presence of sufficient moisture, these reactions result in expansions, which cause extensive cracking in the concrete, particularly when the expansions are unrestrained. Worldwide, AAR leads to various forms of structural damage and is one of the main processes which affect the durability of concrete structures, reducing the service life of these structures and increasing the costs of maintenance of such structures. In Canada and other countries, there is widespread concern about the highway infrastructure, where AAR is one major cause of damage to structures (Kabir, 2006). Due to increased traffic volumes which make bridges more susceptible than ever to deterioration, almost all bridges fail to fulfil the life their design life. Figure 1 is a world map showing different countries in the world where AAR has been encountered and/or reported. This book will cover only a selection of the countries shown in the map in Figure 1 and not all the countries and regions where AAR has been reported.

There are three main types of AAR, distinguishable by the aggregate source. These are: Alkali-Silica Reaction (ASR), Alkali-Silicate Reaction and Alkali-Carbonate Rock Reaction (ACR) (Blight & Alexander, 2011). In order for AAR damage to occur in concrete, four conditions need to be satisfied. These are (Godart & de Rooij, 2013):

- The reactive constituent in the aggregate combination should be present
- The concrete pore water should have sufficiently high alkalinity
- There should be sufficient moisture in the concrete and the surrounding environment
- Portlandite should be present in a sufficient concentration. This condition is specific to ACR, where in the presence of portlandite, the alkali carbonate can transform to calcite and in the process regenerate the alkalis (Godart & de Rooij, 2013). This will be expanded on in later sections in this book.

Figure 1:Countries or areas in the world where AAR in concrete structures has been identified, investigated, reported or where avoidance specification are in place (Poole, 2017)

To prevent or reduce AAR, at least one of the conditions above needs to be eliminated, except in the case of ASR, where one or more of the first three conditions should be eliminated. Therefore, all AAR avoidance or preventive measures will address the elimination of one of the conditions which need to be present for AAR to occur. AAR avoidance measures employed all over the world will be discussed in this book. These measures include those that are practically being applied around the world as well as those that have been proven through experiments to reduce or eliminate AAR in concrete.

Various test methods are employed to test for AAR in concrete. These tests are grouped in different categories, depending on their aim, namely: initial non-quantitative screening tests; indicator tests, performance tests and assessment of occurrence and extent of AAR in existing concrete structures. Many test methods, such as the widely used RILEM procedure, incorporate a combination of different categories in their assessment methodologies. This book will compile a variety of different test methods from all over the world and provide a critical analysis of the different testing methods.

1.3 Effects of AAR

1.3.1 Effects on Serviceability

The most typical manifestation of AAR is concrete cracking. AAR and more particularly ASR results in the formation of gel, which swells and causes expansion within the concrete when there is a presence of water. These internal expansions exceed the tensile capacity of the concrete and result in cracking (Alexander, 2019). The effects of AAR can appear as very unsightly and at other times alarming cracks that can be centimetres wide. AAR may also result in external features such a gel being discharged from cracks, unsightly straining as well as random or map cracking. Figure 2 (Alexander, 2019) shows such cracking in two different structures – one where the cracking is moderate and random and the other where the cracking is severe.

(a)   (b)

Figure 2: (a) Moderate AAR cracking in cope wall and (b) More severe AAR cracking gel discharge in retaining wall (Alexander, 2019)

Furthermore, in some areas in the United States, small pop-outs have been observed on slab surfaces, and these were related to shale aggregates which contain an opaline component. In the Vancouver area (British Columbia), some aggregates which contain iron compounds experience AAR when they are used in combination with high-alkali cements (Alexander, 2019). The reaction results in aesthetic problems such as brown stains and pop-outs.

1.3.2 Effects on Structural Integrity and Performance

When considering structural effects of AAR, it is important to take into account the fact that the expansions which develop in structures due to AAR are vary due to differences in moisture conditions, porosity, alkali concentration as well as the degree of carbonation between the internal part of the concrete and its surface (Oberholster, 2009). Table 1 summarises the effects of ASR on the mechanical properties of concrete.

The data in the Table 1 was summarised by Clark (1989) and it was obtained from tests conducted on concrete cubes, cylinders and prisms which were cast, as well as from cores that were extracted from concrete structures in the United Kingdom. It was noted that the uniaxial compressive strength which was obtained from long specimens such as cylinders or cores (where the length: diameter ratio was equal to or greater than 2.5) is reduced to a greater extent than that obtained from a cube.

Table 1: Lower bound residual mechanical properties as a percentage of values for unaffected concrete at 28 days (Oberholster, 2009)

Property	Percentage strength as compared with unaffected concrete for free expansion (microstrain) as indicated				
	500	1 000	2 500	5 000	10 000
Compressive strength*	100	85	80	75	70
Uniaxial compressive strength **	95	80	60	60	-
Tensile strength	85	75	55	50	-
Elastic modulus	100	70	50	35	30
*Cube **Core, length: diameter ratio = 2.5 or greater					

The uniaxial compressive strength is required for the structural assessment. The residual tensile strength obtained is affected by the method of testing. The tensile strength values in Table 1 is therefore appropriate to the splitting tensile or torsional tensile strength (Oberholster, 2009). The residual strengths and stiffnesses on actual concrete structures are a modification of the figures shown in Table 1. This is due to the fact that concrete in actual structures is restrained by the adjacent material and is therefore in a state of biaxial or triaxial stress (Oberholster, 2009). The impact of this is a reduction of the damage caused to the concrete and an increase in the residual mechanical properties. Additionally, the actual strength of concrete usually exceeds the 28-day design strength by an amount that is greater than the strength reductions resulting from AAR damage. The reduction in compressive strength due to AAR is therefore usually not seen to be a big problem in practice. The reduction in tensile strength is however significant and is worthy of consideration and attention. An important point that should also be taken into account when assessing results such as those in Table 1 is the fact that severely microscopically cracked cores retrieved from structures are normally not used in strength testing. Due to this, the core results may give results which are better than the reality (Oberholster, 2009).

Even though AAR in isolation may not have a detrimental impact on the compressive strength of a structure, it is very important to take into account that the damage it causes with the possible severe map cracking creates a pathway for other deterioration mechanisms such as corrosion to cause major reduction of the compressive strength of concrete. The map cracking may also significantly affect the concrete tensile strength and subsequent bending and deflection of concrete members.

1.3.3 Cost of Infrastructure Management of AAR-Affected Structures

Infrastructure damage due to AAR results in structures failing to serve out their full design life. In Canada, for example, where damage to highway infrastructure due to AAR is excessive, the costs to repair or replace highway bridges are extremely high, and these cause significant challenges to local, provincial and federal governments when having to allocate part of their limited funding to the maintenance and repair of these transportation infrastructures (Kabir, 2006). Oberholster (2009) also recognised that the impacts of concrete deterioration on the durability of concrete due to AAR has major cost implications for the rehabilitation and replacement of structures which have been affected. Also, there are costs involved in employing precautionary measures against the occurrence of AAR in new structures that are being constructed. Exact costs of maintenance, repair and/or rehabilitation of structures specifically due to AAR damage are difficult to estimate because AAR damage usually does not occur in isolation but it normally enables other damage mechanisms to occur at an accelerated rate, and the maintenance of such structures includes damage from mechanisms such as corrosion.

1.4   Aim of Study

This book aims to outline the types of AAR and the mechanisms associated with them, and to highlight case studies of AAR incidences around the world. This book further aims to provide a comprehensive compilation and analysis of various AAR avoidance measures as well as AAR tests that are performed worldwide. Commonalities and differences will be highlighted between the different case studies, and critical analyses will be done on the AAR avoidance measures and AAR tests that will be discussed.

1.5   Study Objectives

The following objectives have been set for this study:
1   Define and give the background on AAR worldwide.
2   Study different types of AAR damage worldwide.
3   Study different cases of AAR incidences around the world and highlight any similarities and/or differences.
4   Study measures employed to avoid AAR worldwide and provide critique on these measures.
5   Study tests used to conduct AAR assessments and provide critique on these tests.

1.6   Study Assumptions

The following assumptions are made for this study:
- Due to the fact that Alkali-Silica reaction (ASR) is the most common type of Alkali Aggregate Reaction (AAR), this book will focus mainly on ASR and not on Alkali-silicate reaction and Alkali-Carbonate Rock Reaction (ACR). Therefore, AAR and ASR will be used interchangeably in this book.
- This study is based on AAR literature from around the world. No additional tests will be performed in this research. Only the information from the literature will be used.

- The author does not undertake to perform any tests to verify any work or results that were obtained by authors of the literature referred in this book.

1.7   Study Outputs

- To perform a study of different cases of AAR damage from around the world.
- To highlight any similarities and/or differences in the different cases of AAR damage around the world.
- To perform a study of the measures employed to avoid AAR worldwide.
- To provide a critical analysis of AAR avoidance and control measures which are employed worldwide.
- To produce a compilation of the tests performed to assess AAR.
- To provide a critical analysis of the different AAR tests performed worldwide.

1.8   Outline of Book

This book comprises six chapters, which are laid out as follows:

Chapter 1 presents the overall introduction of the study, outlining the background, aim, objectives, assumptions and the research outputs.

Chapter 2 discusses the different types of AAR and their governing mechanisms, and also studies cases of AAR that have been encountered around the world, with a highlight of the visual/identification characteristics.

Chapter 3 discusses avoidance measures which have been employed around the world as well as proposed avoidance methods which have been proven experimentally to have potential to prevent or minimise AAR. A critical analysis of these methods is done at the end of the chapter.

Chapter 4 details different tests which are employed to assess AAR around the world. A critical analysis of these test is done at the end of the chapter.

Chapter 5 summarises and concludes the study.

Chapter 6 makes recommendations and suggestions for further work.

2. Types, Mechanisms and Incidences of AAR Worldwide

2.1 Introduction

AAR is divided into three main types, namely Alkali Silica Reaction (ASR), Alkali-Silicate Reaction and Alkali-Carbonate Rock Reaction (ACR). The different AAR types have different variations of the mechanisms which govern their occurrence. ASR is the most common type of AAR and will therefore be the main focus of this book. Many incidences of AAR damage have been reported around the world. This chapter will define the different types of AAR and describe their mechanisms. Various incidences of AAR around the world will then be explored and studied, and the visual/identification characteristics of the AAR damage are highlighted.

2.2 Types of AAR and Associated Mechanisms

There are three main types of AAR, distinguishable by the aggregate source. These are: Alkali-Silica Reaction (ASR), Alkali-Silicate Reaction and Alkali-Carbonate Rock Reaction (ACR) (Blight & Alexander, 2011). These are briefly described below:

2.2.1 Types of AAR
2.2.1.1   Alkali-Silica Reaction (ASR)

ASR is the most common type of AAR. It occurs between the alkalis found in the concrete pore solution and certain reactive forms of silica which are chemically unstable, referred to as alkali-susceptible. Examples of rocks containing such forms of silica include greywacke, quartzite hornfels, phyllite, argillite, granite, granite-gneiss and granodiorite. In ASR, an expansive alkali-silica gel is produced in the concrete. That reaction is governed by Equation 2.1 (Blight & Alexander, 2011):

$$2(\text{Na/K})\text{OH} \quad + \quad \text{SiO}_2 \quad + \quad \text{H}_2\text{O} \quad \rightarrow \quad \text{Na}_2\text{SiO}_3 \cdot 2\text{H}_2\text{O} \tag{2.1}$$

Alkali   Silica   Water   Alkali-Silica gel

The reaction between the alkalis found in the concrete pore solution and the reactive forms of silica found in the aggregates results in the formation of an alkali silica gel. This gel then absorbs water and swells. The swelling action of the gel exerts tensile stress, resulting in concrete cracking if it exceeds the tensile strength of the concrete by about 4MPa (Oberholster, 2009). In order for expansive ASR to take place, the following factors all need to be present:

A sufficient alkali content in the concrete pore solution

Alkalis in cement are typically referred to in terms of the sodium oxide equivalent [%Na_2O-eq = %Na_2O + (0.658 x % K_2O)], which is commonly expressed as a percentage by mass of cement. When the volumetric mass of the concrete is considered (which is more accurate), then the alkali content is expressed in terms of kilograms per cubic metre of concrete (Na_2O-eq/m^3)

(Oberholster, 2009). Alkalis are usually sourced from the cement where they exist as neutral sulphate such as Na_2SO_4, K_2SO_4 or the mixed salt $(Na,K)_2KO_4$. All these are highly soluble and once a high-alkali cement is mixed, the concrete pore solution contains almost entirely Na^+, K^+ and OH^- ions and the pH of the pore solution may be in the range of 13-14. When there is low alkalinity in the concrete pore solution, the pH is usually in the range of 12.5-12.9 (Oberholster, 2009).

Alkalis may also be sourced from the reaction of calcium hydroxide during the hydration of cement with alkali-containing minerals in the aggregates, or it may be sourced from external sources such as mixing water, sea water, sea spray or aggregates which contain salts or from chemical admixtures like sodium lignosulphonate (Oberholster, 2009). A cement is considered a high-alkali cement when the Na_2O-eq is greater than 0.6%, and can be considered unsuitable to use in combination with a potentially highly reactive aggregate in the concrete mix (Oberholster, 2009).

Generally though, the alkalinity of the pore solution is determined by the product of the alkali content of the cement and the amount of cement in the concrete, and this is what determines the extent the alkalis will react with the aggregates. It is therefore more accurate to limit the alkali content per m^3 of concrete and not to the cement content (Oberholster, 2009). Due to this, cement-rich concrete mixes will contain higher contents of alkali per m^3 of concrete and will thus be more likely to cause expansion if used in combination with a reactive aggregates.

Aggregate with a sufficient content of deleteriously reactive minerals
It is possible for either coarse or fine aggregate or a combination of both to be reactive in the concrete mix. In order for alkali silica reaction to occur and for damaging expansion to take place, the aggregate needs to contain a sufficient alkali-reactive constituent and it should also be dense (Oberholster, 2009). Porous aggregates usually have sufficient voids within them to accommodate the gel which forms from the reaction. The expansiveness is influenced by the content of reactive constituent as well as the reactivity of the aggregate. As an example, 2% of opal in sand can result in deleterious expansion, while a minimum of about 20% of Malmesbury metasediment (hornfels, greywacke and phyllite) would be required in sand for it to cause deleterious expansion (Oberholster, 2009).

Around the world, different aggregates have been identified through their service records and laboratory testing as potentially deleterious. In South Africa for example, a list of minerals and rocks was compiled and modified over time particularly for South Africa, showing minerals and rocks which are potentially alkali-aggregate reactive, as shown in Table 2. Similar lists have been compiled in different regions and countries around the world.

Table 2: Minerals, rocks and other substances that are potentially deleteriously reactive with alkalis in concrete (Oberholster, 2009).

Minerals		
Opal Tridymite Cristobalite Chalcedony, cryptocrystalline, microcrystalline or glassy quartz Coarse-grained quartz that is intensely fractured, granulated and strained internally or rich in secondary inclusions Siliceous, intermediate and basic volcanic glasses Vein quartz		
Rocks		
	Rock	**Reactive Component**
Igneous	Granodiorite Charnockite Granite	Strained quartz; microcrystalline quartz
	Pumice Rhyolite Andesite Dacite Latite Perlite Obsidian Volcanic tuff	Silicic to intermediate silica-rich volcanic glass; devitrified glass; tridymite
	Basalt	Chalcedony; cristobalite; palagonite; basic volcanic glass
Metamorphic	Gneiss Schist	Strained quartz; microcrystalline quartz
	Quartzite	Strained and microcrystalline quartz; chert
	Hornfels Phyllite Cataclasite Mylonite Argillite	Strained quartz; microcrystalline to cryptocrystalline quartz
Sedimentary	Sandstone	Strained and microcrystalline quartz; chert; opal
	Greywacke	Strained and microcrystalline to cryptocrystalline quartz
	Siltstone Shale	Strained and microcrystalline to cryptocrystalline quartz; opal
	Tillite	Strained and microcrystalline to cryptocrystalline quartz
	Chert Flint	Cryptocrystalline quartz; chalcedony; opal
	Diatomite	Opal; cryptocrystalline quartz
	Argillaceous dolomitic limestone and Calcitic dolostone Quartz-bearing argillaceous Calcitic Dolostone	Dolomite; clay minerals exposed by dedolomitization
Other substances		
Synthetic glass; silica gel		

The necessary environmental conditions to support the reaction

For expansion and subsequent cracking to occur due to ASR in concrete, the environmental conditions to which it is exposed need to be conducive. Two environmental conditions play the biggest role in the occurrence of ASR, namely: temperature and moisture.

Temperature

In a laboratory, the expansion rate increases with an increase in temperature to about 60°C. In the field, the expansion rate doubles for every 10°C increase in mean annual ambient temperature (Oberholster, 2009).

Moisture

The mixing water content in concrete is usually enough for a reaction to take place and for it to be maintained, unless the concrete loses water and dries out. Usually, ASR will take place when the internal relative humidity is above about 85% (Oberholster, 2009). Any environmental condition that provides or maintains enough moisture so that the internal relative humidity is above the critical levels will therefore aid in the occurrence of ASR.

2.2.1.2 Alkali-Silicate Reaction

Alkali-silicate reaction involves reactions with aggregates such as phyllites, argillites and some greywackes which contain phyllosilicates such as vermiculite, chlorite and mica (Blight & Alexander, 2011). This reaction has been observed to occur when the concrete contains aggregates such as greywacke and argillite found in Nova Scotia in Canada (Oberholster, 2009). In the case of alkali silicate reaction, the reaction is with silica in the combined form with phyllosilicates, as opposed to the reaction with free silica, as is with ASR. Alkali silicate reactions are complicated to characterise but they may result in expansion.

2.2.1.3 Alkali-Carbonate Rock Reaction (ACR)

In ACR, no gel is produced. There is however an expansion of coarse aggregate particles which results from the reaction between alkali hydroxide and small dolomite crystals in a clay matrix (Blight & Alexander, 2011). This then results in a dedolomitization reaction (a reaction in which dolomite is broken down into brucite and alkali carbonate) as shown by Equation 2.2 (Blight & Alexander, 2011):

$$2(Na/K)OH + CaMg(CO_3)_2 \rightarrow CaCO_3 + Mg(OH)_2 + (Na/K)_2CO_3 \qquad (2.2)$$

Alkali   Dolomite   Calcite   Brucite   Alkali carbonate

ACR is limited to carbonate aggregates such as some argillaceous dolomitic limestones which contain clay and dolomite. ACR is not common, and occurs mainly in Canada where the alkali-susceptible carbonate rocks are found in Southern Ontario and in the Ottawa-St Lawrence region (Blight & Alexander, 2011).

2.2.2   Mechanisms of AAR

This section gives a brief description of the mechanisms governing the different types of AAR. Very little literature exists about alkali-silicate reaction and therefore no mechanisms are discussed for this type of AAR. The mechanisms for alkali-silica reaction and alkali carbonate rock reaction are outlined in this section.

2.2.2.1 Alkali-Silica Reaction

Mindess, et al. (2003) reviewed the mechanism of alkali-silica reaction that was described by Helmuth and Stark (1992). This mechanism is now briefly discussed.

It was observed that when alkali-silica reaction takes place, two gel components are formed. One of the components is a non-swelling calcium-alkali-silicate hydrate (C-N (K)-S-H), while the other component is a swelling alkali-silicate-hydrate (N(K)-S-H). A certain amount of non-swelling calcium-alkali-silicate hydrate (C-N (K)-S-H) is always formed when alkali-silica reaction takes place in concrete. This reaction will be safe if this non-swelling gel is the only one which is formed. However, if both the non-swelling and the swelling gels are formed, then the reaction is not safe. The relative amounts of alkali and reactive silica are very important factors to consider in this reaction. The steps involved in this process are as follows (Alexander, 2019):

1. In a pore solution which consists of water and the ions Na^+, K^+, Ca^{2+}, OH^- and $H_3SiO_4^-$ (the $H_3SiO_4^-$ ions result from dissolved silica) the reactive silica goes through depolymerisation, dissolution and swelling. Even though this swelling may result in damage to the concrete, the most significant damage results from the expansion of the reaction products which follows thereafter.

2. Non-swelling calcium-alkali-silicate hydrate (C-N (K)-S-H) is formed from the diffusion of alkali and calcium ions into the swollen aggregate. Due to the fact that the solubility of CH is inversely proportional to the concentration of alkali, the amount of calcium depends on the concentration of alkali.

3. The pore solution diffuses through the porous layer of (C-N (K)-S-H) gel to the silica. The result of this diffusion may be safe or unsafe, depending on the relative concentration of alkali and the rate at which the diffusion takes place. If the (C-N (K)-S-H) is made up of 53% or more CaO on an anhydrous (with no water) weight basis of the gel, then only the non-swelling gel will be formed. However, at high alkali concentration, the solubility of CH is depressed and this results in the formation of swelling (N (K)-S-H) gel, which contains little to no calcium. The (N (K)-S-H) gel has a very low viscosity, which makes it easy for it to diffuse away from the aggregate. However, the presence of (C-N (K)-S-H) results in the formation of a composite gel with a very high viscosity and a low porosity.

4. Through the process of osmosis, the (N (K)-S-H) gel attracts water. Due to this, there is an increase in volume, and local tensile stresses are induced in the concrete. Eventually, cracking takes place in the concrete. The reaction product gradually flows under pressure from the place where it is initially formed and it fills the cracks.

2.2.2.2 Alkali-Carbonate Rock Reaction (ACR)

Swenson (1957), followed by Gillot (1964) and then followed again by Dunkan (1973) described that when certain carbonic rocks from Kingston in Ontario in Canada were used in concrete, they resulted in expansion. It was also noted that the process by which this destructive expansion occurred differed from the process by which the better known alkali-silica reaction occurs (Stefan, 2012). ACR was characterized by the decay of dolomite and the effect of its reaction with alkalis. Due to the fact that ACR is a rare reaction, there has historically been some controversy around the exact mechanisms by which the reaction takes place. Many hypotheses on the process of ACR have been created, and most scientists are of the belief that the process is as follows (Stefan, 2012):

1. Dolomite undergoes dissolution when it reacts with alkalis and produces calcite, brucite and alkali carbonate, as per Equation 2.3 (Stefan, 2012):

$$CaMg(CO_3)_2 + 2MOH \rightarrow Mg(OH)_2 + CaCO_3 + M_2CO_3 \qquad (2.3)$$

   Dolomite   Alkali   Brucite Calcite Thermonitrate

   Where M = K, Na or Li

2. When this reaction takes place in concrete, the natron (which is a mixture of sodium carbonate decahydrate and sodium bicarbonate along with small quantities of sodium chloride and sodium sulfate) can then react with lime ($Ca(OH)_2$), producing sodium hydroxide (NaOH) and calcite ($CaCO_3$), as per Equation 2.4 (Stefan, 2012):

$$M_2CO_3 + Ca(OH)_2 \rightarrow 2MOH + CaCO_3 \qquad (2.4)$$

   Where M = K, Na or Li

   This leads to the recovery of alkalis and the further dissolution of dolomite.

3. In theory, the reactions above proceed until dolomite is completely dissolved. Practically, the products of the dedolomitization reaction may react with other components of the concrete. For example, brucite $Mg(OH)_2$ may react with silica to for hydrated sodium-calcium silica. It is for this reason that there is no silica in the reactive zones of non-dolomitic limestones.

4. It is these mechanisms that are responsible for initiating the expansion process which results in the destruction of the concrete.

2.3 Incidences of AAR Worldwide

This section highlights incidences of AAR that have been encountered worldwide, grouped by geographic locations. Figure 1 in Chapter 1 showed different countries in the world where AAR has been encountered and/or reported. The focus of the section is to highlight some different parts of the world where AAR has been observed, as well as to highlight the identification characteristics of AAR. Structures will be studied from the following geographic areas/regions: The United Kingdom, Russia, North America, India, Southern and Central Africa, Australia, Nordic Europe, Asia, Mainland Europe and North Africa. Where information is available, the overall damage experienced by the different structures will be discussed and the rates at which the AAR took place within those structures will also be highlighted. The section also aims to point out the aggregates used in the construction of the structures to eventually determine if there are any commonalities in the types of aggregates used and the occurrence of AAR in different regions. Commonalities and differences in the AAR experienced in different parts of the world will be drawn, where applicable.

2.3.1 AAR in the United Kingdom

Val de la Mare Dam- United Kingdom

The Val de la Mare Dam is a mass concrete gravity dam located in St Ouen's in Jersey in the United Kingdom. It was constructed from 1957 to 1961, with the dimensions of 29 m height and 168 m length, as shown in Figure 3 (Sims, 2017). The coarse aggregate used in the construction was mainly crushed diorite and granodiorite, combined with Jersey shale, chert and some igneous rocks. The fine aggregate used was a fossil beach sand that consisted of quartzite, quartz and feldspar as well as small contents of chert, shell and igneous rocks.

Figure 3: The Val de la Mare Dam in the United Kingdom in April 1989 (Sims, 2017)

In 1971 – about 10 years after construction - four of the concrete sections exhibited discoloration, dampening and cracking, and there was a displacement of up to 13 mm in some sections, as can be seen in Figure 4 (Sims, 2017). Over the 13 years that followed, the dam was monitored and a continued expansion of 50-75 microstrain per year due to ASR was observed.

Figure 4: Concrete parapet of the Val de la Mare Dam in the United Kingdom with a displacement of up to 13mm (Sims, 2017)

Electricity Substation - United Kingdom

One case of ASR in the UK mainland was an electricity substation at Milehouse in Plymouth, Devon (Sims, 2017). The affected structures were unreinforced concrete bases that were in the ground and had surfaces which were exposed to weather conditions. In 1977, they exhibited random cracking patterns as can be seen in Figure 5 (a) and Figure 5 (b) (Sims, 2017).

The concrete used in the construction consisted of the following constituents (Sims, 2017):
- coarse aggregate: crushed limestone,
- fine aggregate: sea-dredged aggregate containing a subordinate proportion of chert,
- cement: locally produced high- alkali cement (approximately 1 -1.2% Na_2Oeq)

The ASR could be observed through brown and white chert particles, as well as reaction rims which formed in the concrete. Gel deposits were also observed. The surfaces of the bases were exposed to wetting and drying, and this was a necessary condition in promoting the occurrence of ASR (Sims, 2017).

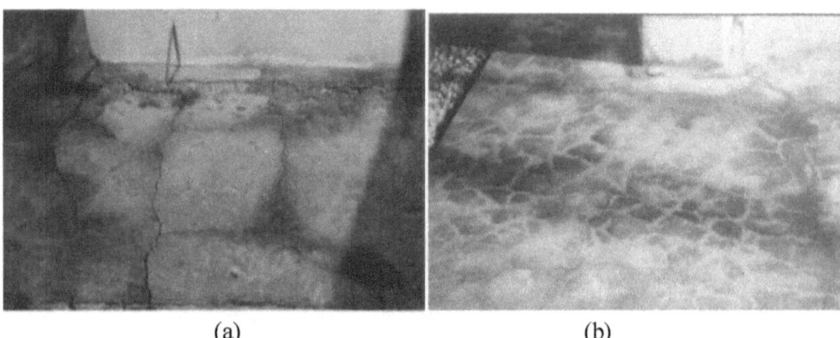

Figure 5:(a) and (b) Map cracking on concrete bases at Milehouse electricity substation in Plymouth in 1977 (Sims, 2017)

Car Park - United Kingdom

One case of ASR occurrence was the Charles Cross Multi-storey Car Park in Plymouth. It was built in 1970 and started to suffer slight to severe cracks just 7 years after its construction. Severe superficial cracks were observed in some beams, while some column heads exhibited severe cracking and fragmentation, as can be seen in Figure 6 (Sims, 2017). Map cracking patterns were observed on the ground-level retaining walls, as well as some of the parapet walls and pavements on the top floor.

It was shown through laboratory tests that the concrete constituents used in the construction of the car park consisted of (Sims, 2017):

- coarse aggregate: non-reactive crushed limestone or crushed granite,
- fine aggregate: sea-dredged aggregate containing reactive chert particles,
- cement: high- alkali Portland cement

Both cases above had similar concrete constituents. Over time, it was becoming clear that the main cause of damaging ASR in the UK was the combination of fine aggregate which contained chert (or flint), a coarse aggregate such as crushed limestone which was non-reactive and a high-alkali Portland cement (Sims, 2017).

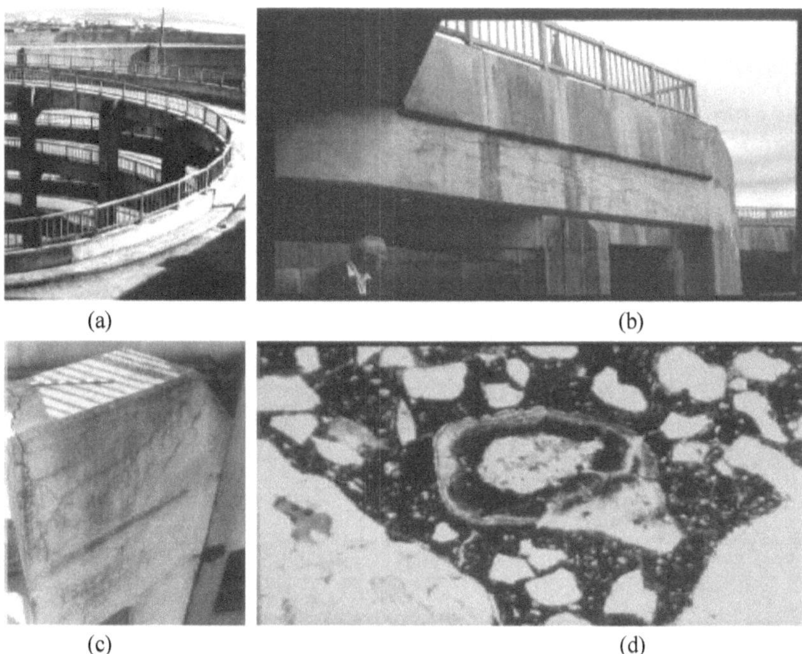

Figure 6: (a) The spiral ramp of the Charles Cross Car Park in Plymouth; (b) Severely cracked beam; (c) Cracked column head; (d) Photomicrograph of a thin section of concrete from the car park (Sims, 2017)

Commonalities and differences in the AAR in The United Kingdom
In the structures studied in The United Kingdom, the type of structures affected by AAR included dams, substations and car parks. Some common aggregates that were observed between the different structures were crushed limestone and aggregated containing chert. Other aggregates were granites, quartzites and feldspars. It was also seen that for the structures studied in The UK, the damage due to AAR can occur as early as 7 years from the time of construction. Map cracking and reaction rims around aggregates are very common characteristics in the structures studied in The UK. The water retaining structure studied, i.e. the Val de la Mare Dam, exhibited displacement of the dam walls. Table 3 further draws out the commonalities and differences between the structures in The UK.

Table 3: Comparison of AAR in the United Kingdom

Structure	Country	Aggregates used	AAR characteristics	Time from construction to AAR damage	Overall damage
Val de la Mare Dam	United Kingdom	Crushed diorite, granodiorite, Jersey shale, chert, igneous rocks, quartzite, quartz and feldspar	Discoloration, dampening and cracking, expansion of 50-75 microstrain	10 years	Displacement of up to 13 mm
Electricity Substation	United Kingdom	Crushed limestone, sea-dredged aggregate containing chert	Random cracking, brown and white chert particles, reaction rims in concrete, gel deposits	N/A	Severe cracking and deterioration of substation bases.
Car Park	United Kingdom	Crushed limestone, crushed granite, chert	Slight to severe map cracking	7 years	Substantial cracking and fragmentation of some columns

2.3.2 AAR in Russia

Railway Sleepers – Russia

Many cases of alkali silica reaction damage have been recorded in Russia. The most notable signs of concrete damage due to ASR includes crack networks and white jelly-like extrusions from the concrete. The commonly used aggregates are flint, sandstone and andesites. The most well-known cases are of house footing structures, railway sleepers, port facilities as well as concrete elements in public and industrial buildings and structures (Falikman & Rozentahl, 2017). It was reported in the year 2004 that there was large-scale deterioration of precast concrete railway sleepers which were produced in 2001. This deterioration manifested through cracks with varying widths as can be seen in Figure 7. Core samples were extracted from the damaged sleepers and it could be seen that the deteriorated coarse aggregate grains released a gel, which is one of the characteristics of alkali-silica reaction (Falikman & Rozentahl, 2017). Similar cases of ASR damage reported in Russia included the failure of footings of a transmission tower of the USSR railway overhead system, which took place 3 years after construction (Falikman & Rozentahl, 2017).

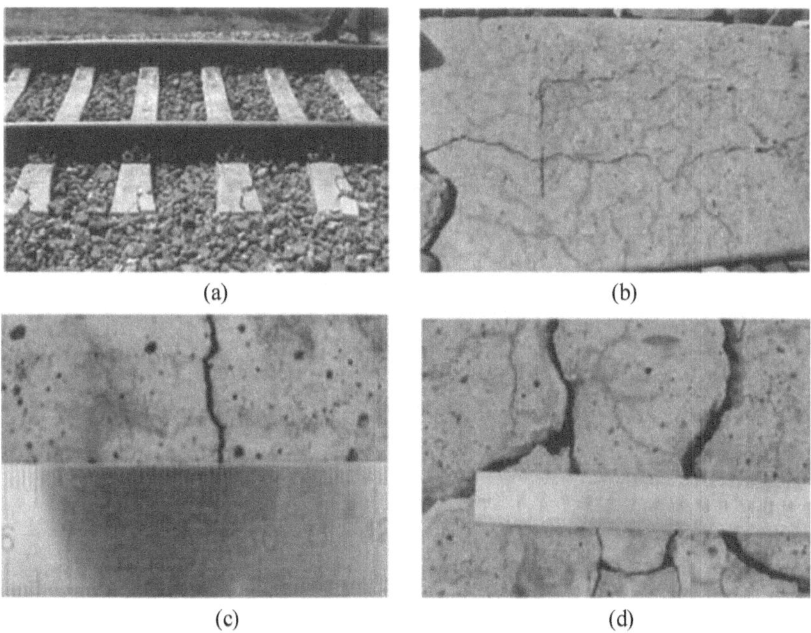

(a)   (b)

(c)   (d)

Figure 7: (a) and (b) General view of the deteriorated railway sleepers; (c) and (d) Close up view of the cracks in the railway sleepers (Falikman & Rozentahl, 2017)

Summary of AAR in Russia

Table 4 shows a summary of the details of the AAR-affected structure studied in Russia.

Table 4: Summary of AAR in the Russia

Structure	Country	Aggregates used	AAR characteristics	Time from construction to AAR damage	Overall damage
Railway Sleepers	Russia	Flint, sandstone and andesites	Cracks with varying widths deposits on the concrete surface, gel	3 years	Disintegration of railway sleepers and crack widths up to 1 cm

2.3.3 AAR in North America

Bridge Decks – Canada

AAR is one of the most common causes of infrastructure deterioration in Eastern Canada. There has been a rise of unprecedented concrete deterioration which has caused widespread concern about the highway infrastructure in Canada (Kabir, 2006). Figure 8 shows the AAR damaged bridges in the St Lambert Lock Bridge in Montreal and a train bridge located in Sherbrooke. AAR results in swelling and eventual cracking in concrete, and the level of swelling gives an indication of the concrete deterioration, loss of rigidity and decreased mechanical properties. Components obtained from the St. Lambert Lock Bridge were severely affected by AAR, and exhibited various rates of concrete swelling. Components retrieved from bridges in Sherbrooke, including the train-bridge, all exhibited varying degrees of concrete deterioration.

(a)   (b)   (c)

Figure 8: (a) and (b) show AAR induced map cracking in the bridge component at the St. Lambert Lock in Montreal. (c) is the train-bridge beam in Sherbrooke. Both bridges are in Canada (Kabir, 2006)

Parker Dam – United States of America

The Parker Dam between Arizona and California was constructed between 1934 and 1938. In 1940, the United States Bureau made a diagnosis that the expansion and cracking which were observed in the dam were the result of ASR (Thomas, Folliard, & Ideker, 2017). The dam was one of the first large structures in the world to have been diagnosed with ASR and remains in service to date. Two to three years after construction of the dam came to an end, cracks were observed on the dam and they were said to be due to the reaction between the cement and aggregate, which was andesitic aggregate (Thomas et al., 2017). The Parker Dam can be seen in Figure 9.

Cores taken from ASR-affected (where high-alkali cement was used in the construction) areas of the Parker Dam as well as cores taken from non ASR-affected (where low-alkali cement was used in the construction) areas of the same dam were compared to cores and cylinders from the Hoover Dam, also in the United States of America. Cylinders were cast during the construction, while cores were taken at 60 years or later after construction. The results of strength tests performed on these cores are shown in Figure 10 (Thomas et al., 2017).

(a)   (b)

Figure 9: (a) Parker Dam in the USA exhibiting map cracking in 2013; (b) Visible cracking on elements of the Parker Dam (Thomas, Folliard, et al., 2017)

The results in Figure 10 show that the cores taken from the parts of the Parker Dam constructed with low-alkali cement showed no ASR and their strength development was similar to that of the Hoover Dam. The cores taken from the ASR-affected parts of the Parker Dam, which were constructed with high-alkali cement showed a reduction in concrete strength (Thomas et al., 2017).

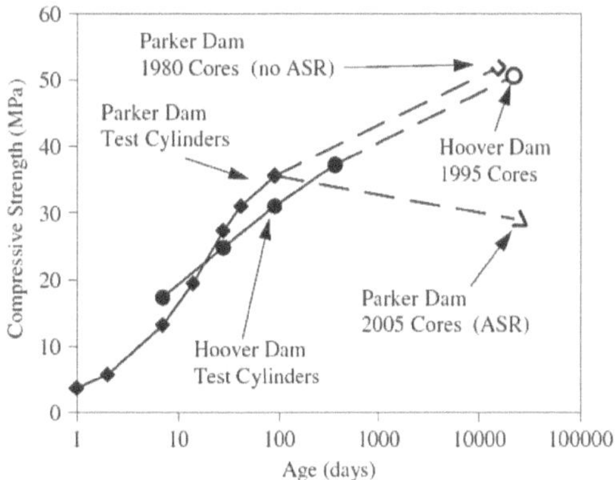

Figure 10: Effect of ASR on the strength of cores extracted from Parker Dam (Thomas, Folliard, et al., 2017)

Mactaquac Dam – Canada

The Mactaquac Generating Station is located in the New Brunswick province, 20 km from the City of Fredericton. It is made up concrete structures which include an intake structure, a powerhouse and two spillways (Thomas, Folliard, et al., 2017). About 10 years after its construction was completed, the concrete structures started to show signs of deterioration, and this was attributed to alkali silica reaction and the expected life of the dam was reduced significantly. The aggregate used in the concrete was greywacke/argillite rock. The dam is shown in Figure 11. Since its construction, the height of the intake structure is said to have increased by more than 175mm (Thomas et al., 2017). Additionally, the unrestrained expansion of the concrete is currently estimated to be in the range of 120 – 150 microstrain per year and there is no indication of a decrease in this rate of expansion. Due to the continued problems in the operations as well as increasing maintenance costs, partially due to ASR damage, the concrete structures are projected to be replaced by the year 2030, at which point they will be in service for 60 years (Thomas et al., 2017).

<table>
<tr><td>(a)</td><td>(b)</td></tr>
</table>

Figure 11: (a)Mactaquac Dam in Canada exhibiting map cracking in 2008; (b) Cracking on dam elements (Thomas, Folliard, et al., 2017)

Commonalities and differences in the AAR in North America

In the structures studied in North America, the type of structures affected by AAR included dams, and bridge decks. The aggregates that were used in the different structures were limestone, andesitic aggregate and greywacke. Other aggregates were granites, quartzites and feldspars. It was also seen that for the structures studied in North America, the damage due to AAR can occur as early as 2 years from the time of construction. Map cracking, concrete swelling and gel exudence are very common characteristics in the structures studied in North America. In these structures, it was seen that all the structures suffered a loss of structural performance. Table 5Table 3 further draws out the commonalities and differences between the structures in North America.

Table 5: Comparison of AAR in North America

Structure	Country	Aggregates used	AAR characteristics	Time from construction to AAR damage	Overall damage
Bridge Decks	Canada	Reactive limestone	Concrete swelling, map cracking	N/A	Concrete cracking, loss of rigidity, decreased mechanical properties
Parker Dam	United States of America	Andesitic aggregate	Map cracking	2 to 3 years	Compressive strength reduction
Mactaquac Dam	Canada	Greywacke/ argillite rock	Map cracking, gel exudence, expansion of 120 – 150 microstrain per year	10 years	Reduction in service life

2.3.4 AAR in India

Hirakud Dam Spillway – India

The Hirakud dam is one of the longest earth dams in the world, at a length of 4.8 km and is composed of earth, concrete and masonry. Investigations were done on spillways in the Hirakud dam 27 years after its construction (Mullick, 2017). The spillways were found to be extensively cracked. Map cracking, superimposed with longitudinal horizontal cracks were observed on the spillways. It was also noted that the extent of the cracks had been increasing with time (Mullick, 2017). Samples retrieved from the structure showed that there was indeed presence of ASR, which manifested through off-white, translucent-to-opaque agglomeration of fluffy gel-type deposits in the voids on the borders with aggregates as well as on the aggregates as can be seen in Figure 12 (Mullick, 2017). The petrographic examination identified that the three coarse aggregates present in the concrete were quartzite river shingles, granitic rocks and diorites (Mullick, 2017).

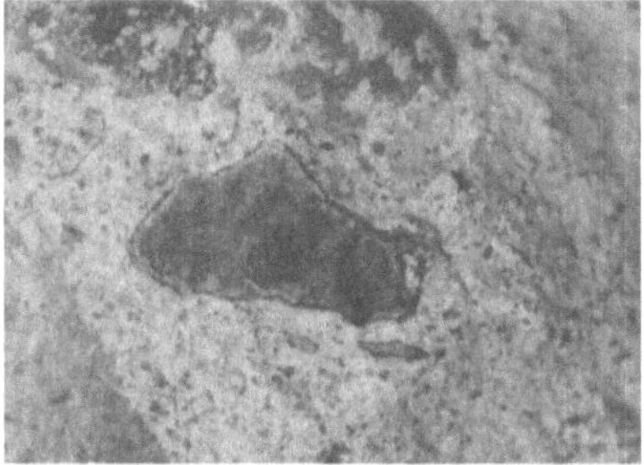

Figure 12: Concrete core sample extracted from the Hirakud dam spillway in India showing a dark reaction rim around aggregates and ASR gel (Mullick, 2017)

Rihand Dam and Powerhouse Structure – India

The Rihand concrete gravity dam and the adjacent powerhouse showed signs of ASR damage 25 years after it was constructed (Mullick, 2017). The ASR damage manifested as cracking of concrete, misalignment of hydro mechanical machinery as well as movements in the concrete which resulted in difficulties with operating the gates, cranes and passenger lifts (Mullick, 2017). Additionally, the powerhouse also experienced several problems with machinery and operations as a result of the concrete movement. Amongst these problems was a horizontal displacement of 30mm between the powerhouse crane girders in different bays, resulting in intake gates not sealing properly. The powerhouse also tilted upstream (Mullick, 2017).

When concrete samples of the broken and damaged surface from the structure were examined through visual inspection and by using a magnifying glass, they were found to contain white deposits in the voids in the concrete and on the aggregates, which is typical to ASR. In other cases, reaction rims around the aggregates were observed, and this is also another manifestation of ASR (Mullick, 2017). A petrographic examination was performed on aggregates which were extracted from the concrete. This revealed that the aggregates used were mainly biotite granite, muscovite granite and mica granite (Mullick, 2017). Each of these rocks contained:

- a quartz content in the range of 32 – 45%
- alkali (sodium potassium and sodium calcium feldspars such as orthoclase, microcline and plagioclase ranging in content from 35 to 45%, and
- varying contents of biotite, muscovite and other minerals including iron ore, chlorite and apatite.

Commonalities and differences in the AAR in India

In the structures studied in India, the type of structures affected by AAR were mainly dam structures. Some common aggregates that were observed between the different structures were granitic rocks. Other aggregates were quartzites and diorites. It was also seen that for the structures studied in India, the damage due to AAR occurred 25 years from the time of construction. Map cracking, gel deposits and white reaction rims around aggregates are very common characteristics in the structures studied in India. In both structure studied, displacement of the concrete was recorded. Table 6 further draws out the commonalities and differences between the structures in India.

Table 6: Comparison of AAR in India

Structure	Country	Aggregates used	AAR characteristics	Time from construction to AAR damage	Overall damage
Hirakud Dam Spillway	India	Quartzite river shingles, granitic rocks and diorites	Map cracking, longitudinal horizontal cracks, gel-type deposits in voids and on aggregates	27 years	Relative horizontal and vertical column displacements of 12 mm and 3 mm respectively due to ASR expansion, shifts in beams, foundation settlement
Rihand Dam and Powerhouse Structure	India	Biotite granite, muscovite granite and mica granite	Cracking of concrete, white deposits in the voids in the concrete and on the aggregates, reaction rims around the aggregates	25 years	Movements in the concrete which resulted in difficulties with operating the gates, cranes and passenger lifts; horizontal displacement of 30mm between the powerhouse crane girders in different bays

2.3.5 AAR in Southern and Central Africa

Reservoirs – Namibia

In Namibia, several reservoirs close to the coastal towns were reported to show signs of ASR. Beushausen (2011) reported on the condition assessments performed on reservoirs in the area. The Collector 2 Reservoir is a 40 year old rectangular reservoir located in the Namib Desert just past Rooibank approximately 25km from the shoreline and at times is subjected to mist from the ocean. The structure was reported to show crack patterns that were consistent with ASR (Beushausen, 2011a). The Mile 7 Square Reservoir is a 30 year old rectangular reservoir located close to Dune 7 just outside the coastal town of Walvis Bay. From an assessment of the exposed fractured concrete surfaces, ASR was clearly visible and was indicated by the white reaction rim around the aggregates as can be seen in Figure 13 (Beushausen, 2011c).

The Mile 7 Round Reservoir is an old decommissioned round reservoir located close to Dune 7 just outside the coastal town of Walvis Bay. A visual assessment was performed on the structure and there was excessive random cracking indicated the possibility of ASR damage (Beushausen, 2011b). The Rooibank Reservoir is a 40 year old rectangular reservoir made up of columns and beams which support arched walls and is located in the Namib Desert just past Rooibank approximately 25km from the shoreline and at times is subjected to mist from the ocean. The structure showed a significant amount of random cracking on all the walls, beams and columns which is consistent with ASR (Beushausen, 2011d).

Figure 13: White reaction rim around aggregates in the Mile 7 Square Reservoir in Namibia (Beushausen, 2011c)

In Namibia, the most alkali-susceptible rocks used are the granite and gneiss (Oberholster, 2009). In the Walvis Bay area, the aggregate commonly used is coarse aggregate from a granite quarry. It was however noted that there are big differences in the reactivity of aggregates from the same geological formations and even between aggregates from the same quarry. Due to this, it is important that aggregate reactivity is established through service record or laboratory testing (Oberholster, 2009)

Nalubaale Power Plant – Uganda, Central Africa
The Nalubaale Power Plant shown in Figure 14 was constructed between 1951 and 1954. ASR damage has now been observed on the concrete structure (Alexander & Blight, 2017). The aggregates used in the structure was the schistose aggregates which were excavated from the site. Expansion and damage due to ASR is expected to continue throughout the life cycle of the plant, but during the remainder of its life, remedial techniques are going to be employed (Alexander & Blight, 2017).

(a)   (b)

Figure 14: (a) The main dam of the Nalubaale Falls and (b) the tailrace and power station at Nalubaale in Uganda (Alexander & Blight, 2017)

Mast Foundations - Namibia

In Namibia, concrete mast foundations at the coastal town of Swakopmund were found to be affected by ASR. The coarse aggregate used in their construction was granite which was sourced from a quarry in Walvis Bay (Oberholster, 2009).

Airport Apron and Mast Footings – Mpumalanga, South Africa

ASR damage in Mpumalanga was observed in an airport apron concrete mast footings. These contain severely cracked granite coarse aggregate (Oberholster, 2009).

Various Structures – Gauteng and Free State, South Africa

Several structures including reservoirs, bridges and an airport runway were reported to be affected by ASR in the Gauteng and Free State region. The ASR was observed in structures were the Witwatersrand Supergroup quartzites and shales were used in the concrete (Alexander & Blight, 2017). In general though, it was noted that the ASR does not continue unless there is an availability of additional moisture due to poor drainage, ponding and poor detailing which allows moisture to accumulate (Alexander & Blight, 2017).

Various Structures – Western Cape, South Africa

Structures such as various bridges, culverts, dams, a hydro-electric power station and sports stadiums have been reported to be affected by ASR in the Western Cape. In the Cape Peninsula and the surrounding areas, the main coarse aggregate used in the constructions are the greywacke rocks of the Malmesbury Group (Alexander & Blight, 2017). Further north, ASR was also reported in prestressed concrete railway sleepers of the Sishen-Saldanha railway line. The aggregate used in that area is granite (Alexander & Blight, 2017).

Various Structures – Eastern Cape, South Africa

In the Eastern Cape, the structures reported to be affected by ASR include irrigation and water storage dams, an irrigation water concrete pipe line, bridges, airport structures, pile caps, retaining wall and lighting mast foundations (Alexander & Blight, 2017).

Motorway Portal Frame - Johannesburg, South Africa

The reinforced concrete motorway portal frame is part of a series which supports a doubled-decker stretch of urban freeway in Johannesburg, the layout of which is as shown in Figure 15. The structure was designed in 1963 and was reported to exhibit signs of severe AAR deterioration (Alexander & Blight, 2017). In 1982 and then again in 1988, full-scale loading tests were performed on the portal frame. Despite the severe damage, the portal frame continued to behave as if the concrete was not cracked (Alexander & Blight, 2017). Overall, it was concluded from the loading tests that the portal frame's strength was adequate for it to continue service, but cosmetic repairs were required to improve visual outward appearance and prevent further widening of existing cracks.

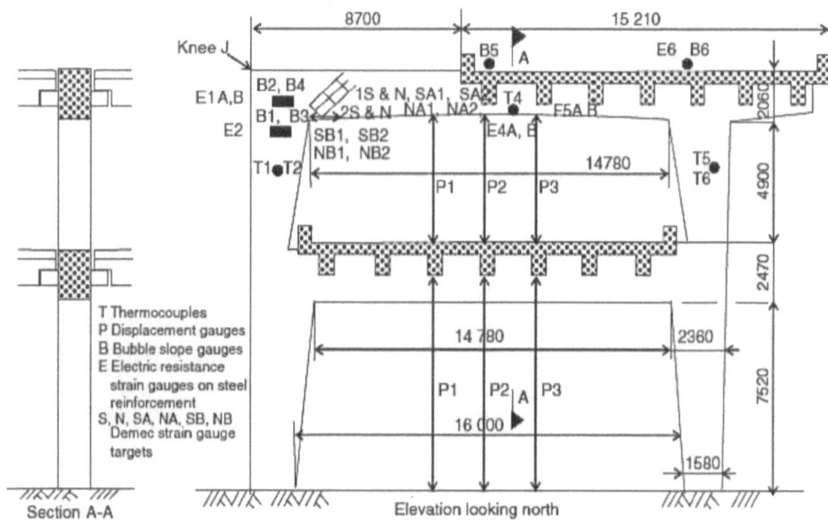

Figure 15: Layout of motorway portal frame in Johannesburg, where loading tests were performed in 1992 and 1998, and subsequent repairs were done. (Dimensions in mm)(Alexander and Blight, 2017)

The repair method which was adopted was to demolish the damaged length of the upper beam of the portal frame, and the reconstruct it with reinforced concrete using the aggregate which is not susceptible to AAR (Alexander & Blight, 2017). The repair was carried out in 1991, during which the replacement beam was constructed with concrete containing dolomite aggregate which is not susceptible to AAR and it also provided to desired stiffness to the concrete. Inspections were performed on the structure in 2003 and it was found that there were very little changes in the cracking and that the repair was still in excellent condition (Alexander & Blight, 2017).

Unreinforced Concrete Road Pavement – Cape Town, South Africa

The 27 km long unreinforced concrete road pavement formed part of a dual carriageway, 4 lane National Route 2 near Cape Town, South Africa, and was constructed in 1969 (Alexander & Blight, 2011). This pavement had deteriorated due to ASR. In 1975, numerous fine hair cracks close to the joints were noticed. The cracks passed through the coarse aggregate, and not around them, and the broken aggregate surfaces also showed reaction rims, which is typical of AAR (Alexander & Blight, 2011).

In 1979, the pavement was evaluated via load testing to investigate the impact of AAR on the remaining pavement life. The test revealed that despite the AAR occurrence, the pavement was still strong and serviceable to carry out its design life (Alexander & Blight, 2011).

Underground Mass Concrete Plug - Free State, South Africa

A reservoir of dimensions 3.5m wide by 3.3m high, at a depth of 2m was constructed in 1981 in a gold mine in the Free State province. The reservoir was a receiver for the compressed air supply in the underground mining. This reservoir was closed with a 1.5m thick unreinforced pressure-retaining concrete plug (Alexander & Blight, 2011). When the concrete plug was designed, no considerations were made for the possibility of AAR, and in 1984, signs of AAR were noticed in the plug. It had a compressor which fed the reservoir through a 250 mm diameter pipe and the air was drawn off through a 500 mm diameter pipe, as shown in Figure 16 (Alexander & Blight, 2011). Many of the conditions necessary for AAR to occur were present. The aggregate used in the concrete was the Witwatersrand quartzite, which is susceptible to AAR, there was a constant supply of free moisture on the plug, and a high and constant temperature. The plug was also constructed with high strength concrete of 50 MPa, which was possibly difficult to mix and place, due to the fact that it was a cement-rich concrete (Alexander & Blight, 2011).

In 1988, a proof test was carried out on the plug, using the dials shown in Figure 16. The plug experienced shrinkage cracking initially, but this was reversed by the expansion due to AAR because the plug was confined on the sides by underground rock walls (Alexander & Blight, 2011). The results of the test showed that the strength and elastic modulus of the concrete did not deteriorate significantly and that even though the plug had expanded due to AAR, it was restrained by the confining rock wall.

Commonalities and differences in the Southern and Central Africa

In the structures studied in Southern Africa and Central Africa, the type of structures affected by AAR included water retaining structures, bridges, power plants, road pavements and underground structures. Some common aggregates that were observed between the different structures were the Witwatersrand Supergroup quartzites and shales, greywackes from the Malmesbury group, granites and gneiss aggregates. Other aggregates were schistone aggregates. It was also seen that for the structures studied in this region, the damage due to AAR can occur as early as 3 years from the time of construction. Map cracking, reaction rims around aggregates and concrete expansion are very common characteristics in the structures studied in this region. The type of overall damage ranged from severe structural damage to no impact of the structural integrity. Table 7 further draws out the commonalities and differences between the structures in The UK.

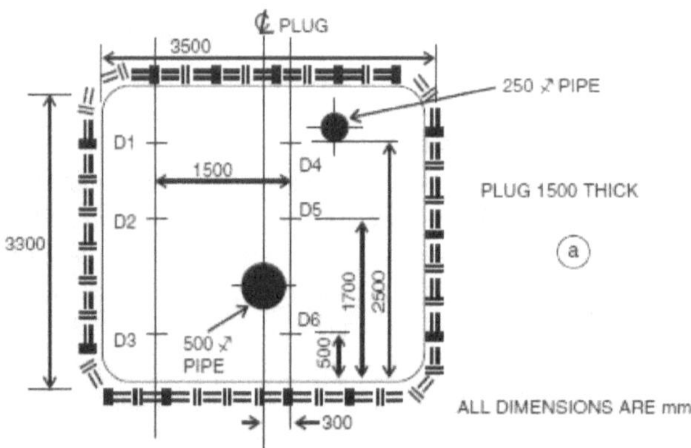

Positions of dial gauges (D1, D2, D3, D4, D5, D6 and
LVDTs (D3 and D5)

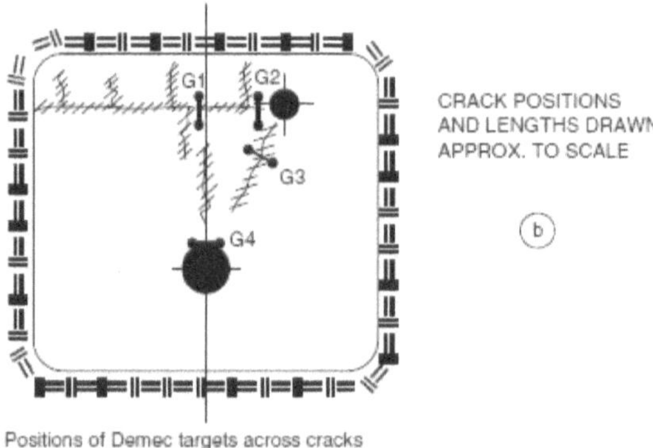

Positions of Demec targets across cracks
(G1 to G4)

Figure 16: (a) and (b) Layout of the concrete plug and location of cracks (Alexander and Blight, 2011)

Table 7: Comparison of AAR in Southern and Central Africa

Structure	Country	Aggregates used	AAR characteristics	Time from construction to AAR damage	Overall damage
Reservoirs	Namibia	Granite and gneiss	Random cracking, white reaction rim around aggregates in	30 years	Extensive concrete deterioration, fractured concrete surfaces
Nalubaale Power Plant	Uganda, Central Africa	Schistose aggregates	Expansion and cracking	40 years	Differential movement between upper and lower galleries
Mast Foundations	Namibia	Archaean Granite and Gneiss	Cracking	N/A	Severe structural deterioration from cracking
An airport apron concrete mast footings	Mpumalanga, South Africa	Archaean Granite and Gneiss	Cracking	N/A	Severe structural deterioration from cracking
Various Structures	Gauteng and Free State, South Africa	Witwatersrand Supergroup quartzites and shales	Map cracking	N/A	N/A
Various Structures	Western Cape, South Africa	greywacke rocks of the Malmesbury Group and granite	Map cracking	N/A	N/A
Various Structures	Eastern Cape, South Africa	N/A	Map cracking	N/A	N/A
Motorway Portal Frame	Johannesburg, South Africa	N/A	Severe Cracking	N/A	Severe cracking but no impact on structural integrity
Unreinforced Concrete Road Pavement	Cape Town, South Africa	N/A	Fine hair cracks passing through the coarse aggregate; broken aggregate surfaces showed reaction rims	6 years	Severe cracking but no impact on structural integrity
Underground Mass Concrete Plug	Free State, South Africa	Witwatersrand quartzite	N/A	3 years	N/A

2.3.6 AAR in Australia

Water Retaining and Other Structures – Australia

AAR generally results in random cracking and map cracking of the affected concrete elements. This usually occurs in mass concrete structures or in lightly reinforced concrete elements. The type of cracking resulting from AAR as shown in Figure 17 (Shayan & Freitag, 2017). The concrete elements shown are a retaining wall in a dam, the end of a crosshead beam in a bridge, a large pylon in a different bridge, as well as columns in a spillway bridge (Shayan & Freitag, 2017). All these concrete elements are part of structures located in South Australia, Victoria and Western Australia. The aggregates commonly found in Victoria are reactive to slowly reactive gneissic quartz gravels (Shayan & Freitag, 2017).

Some other visual manifestations of AAR observed in dams in Australia include the distortion, misalignment and rotation of the AAR-affected elements relative to other elements next to them. Figure 18 shows a concrete block in a dam wall which was displaced and rotated relative to the concrete block next to it (Shayan & Freitag, 2017). The cause of this was the significant amount of expansion and cracking in the one concrete block due to AAR damage.

(a)   (b)

(c)   (d)

Figure 17: (a) AAR induces map cracking in retaining wall of a dam; (b)ASR cracking at the end of a crosshead beam in a bridge; (c) cracking due to ASR in a bridge pylon, (d) a cracked bridge pier in dam spillway, all in Australia (Shayan & Freitag, 2017)

Figure 18: Large concrete block in a dam in Australia displaced (Shayan & Freitag, 2017)

Summary of AAR in Australia
Table 8 shows a summary of the details of the structure studied in Australia.

Table 8: Comparison of AAR in Australia

Structure	Country	Aggregate s used	AAR characteristics	Time from construction to AAR damage	Overall damage
Water Retaining and Other Structures	Australia	Gneissic, quartz gravels	Random and map cracking,	N/A	Displaced and rotated dam wall

2.3.7 AAR in Nordic Europe

Bridges – Denmark

Research conducted in Denmark in the 1960s revealed that opaline flint and calcareous opaline flint were the most prevalent types of reactive aggregates in that country (Lindgard et al., 2017). These reactive compounds were present in both the fine and coarse aggregate sources, including both the land-based and sea-dredged sources. It was found that between of 90 and 95% of the ASR damage in concrete structures was the result of porous opaline or calcareous opaline in the sand fraction of the aggregate (Lindgard et al., 2017). The ASR damage caused by opaline or calcareous opaline flint was observed to occur as rapidly as under 5 years under severe conditions, which was different from other Nordic countries in which only slowly reactive aggregates were used.

In Denmark, ASR was reported in all the different types of moisture exposed outdoor concrete conditions, and it was also reported in swimming pools. Also, severe ASR damage that occurred on many bridges which were built in the 1960s and 1970s were linked to the significant increase in alkali content due to permeation of de-icing salts such as NaCl. The biggest contributing factor to the ASR damage is the fact that during the intrusion of de-icing salts, the successive reaction of highly reactive porous flint opens up the concrete and therefore increases the rate of ingress of alkalis (Lindgard et al., 2017). In addition, the water/cement ratio in several bridges was high. A high water/cement ratio is typically considered to be higher than 0.7 (Alexander & Beushausen, 2009)This has however only been observed to occur on bridges where the bridge deck membranes are defective or have leakages. Due to this, many bridges in Denmark are severely cracked due to ASR, as can be seen in Figure 19 (Lindgard et al., 2017).

Figure 19: Cracking due to ASR on a bridge in Ølstykke in Denmark (Lindgard et al., 2017)

Various Structures – Sweden

ASR was first discovered in Sweden in 1975 in several concrete floors. It manifested as pop-outs caused by porous flint in the south-western part of Sweden. Although ASR in Sweden is not severe, it has been recognized to be fairly common, particularly in structures which were built in the 1960s and 1970s (Lindgard et al., 2017). It was also recognized that some old bridges have undergone ASR, in which the reaction resulted from deicing salts combined with frost action. The most commonly used reactive aggregates in Sweden are found in the South western part of Sweden (in Scania) and they are limestone with fling (chert) and sometimes porous flint (Lindgard et al., 2017).

Many concrete dams and bridges have exhibited ASR damage. ASR damage was also detected on other structures such as balconies, swimming pool complexes and parking decks. The most common manifestation of ASR is pop-outs (Lindgard et al., 2017).

Commonalities and differences in the AAR in Nordic Europe

In the structures studied in Nordic Europe, the type of structures affected by AAR included dams, bridges and concrete floors. Some common aggregates that were observed between the different structures were flint, which was used in both structures studied. It was also seen that for the structures studied in this region, the damage due to AAR can occur as early as 5 years from the time of construction. Map cracking and pop outs are very common characteristics in the structures studied in Nordic Europe. Table 9 further draws out the commonalities and differences between the structures in this region.

Table 9: Comparison of AAR in Nordic Europe

Structure	Country	Aggregates used	AAR characteristics	Time from construction to AAR damage	Overall damage
Bridges	Denmark	Opaline flint and calcareous opaline flint	Map cracking, aggravated by de-icing salts	Under 5 years	Heavy delamination and extensive cracking
Various Structures	Sweden	limestone with flint, porous flint	Pop-outs	Over 10 years	No severe damage

2.3.8 AAR in Asia

Sewage Works – Hong Kong

In Hong Kong, aggregates are normally imported from other places and the reactivity of these aggregates depends on where they were sourced from, as well as the combination in which they are used. The main aggregate used for construction in Hong Kong is local and imported granitic rock, and ASR damage has been reported (Yamada & Miyagawa, 2017). ASR damage by late expansion was also reported on sewage tanks as seen in Figure 20. They were a result of devitrified rhyolitic tuff coarse aggregate sourced from the south of China (Yamada & Miyagawa, 2017).

Concrete Pavement – Korea

On the western coast of South Korea, damage to a concrete pavement at Seohae Expressway was observed and reported. No detailed assessment was performed to determine the type of aggregate used in the concrete (Yamada & Miyagawa, 2017). Judging from the location of the pavement, the major local aggregate type found in the area was Pre-Cambrian (which includes rocks such as gneiss, greenstone, granite, greywacke, iron-formation, schist, basalt, gabbro and anorthosit), with some Palaeozoic rocks (which include shale, limestone, dolomite, sandstone, conglomerate, breccia and chert). There is also the possible presence of rocks with alkali-reaction potential, most likely the cryptocrystalline quartz, which is a late expansive aggregate type, or the chalcedony, which is a more rapidly expansive rock (Yamada & Miyagawa, 2017). In the north-eastern parts of North Korea, rapidly expansive aggregates are expected as a result of Neogene and Quaternary volcanic activities.

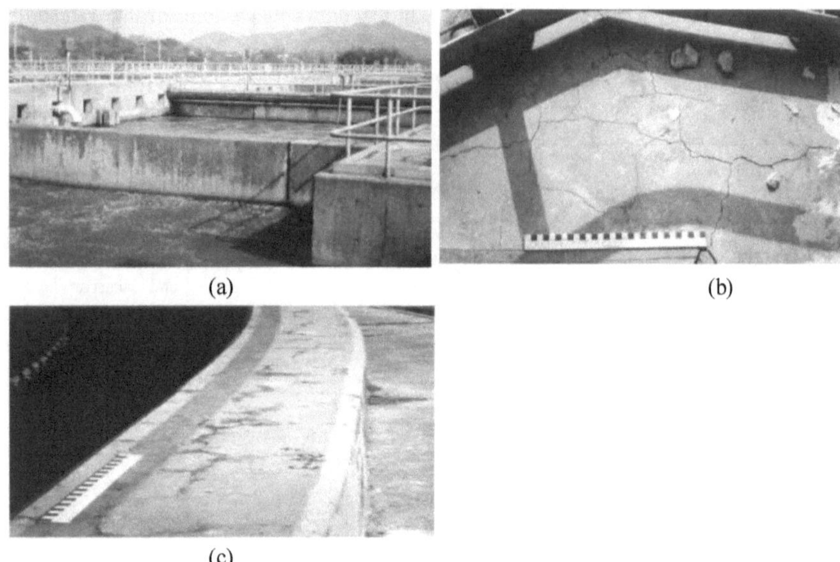

Figure 20: ASR damaged sewage works in Hong Kong. (a) General view; (b) Map cracking in concrete tank wall; (c) Cracking along the top of the surface of concrete wall (Yamada & Miyagawa, 2017)

Commonalities and differences in the AAR in Asia

In the structures studied in Asia, the type of structures affected by AAR included sewage work structures and concrete pavements. Some aggregates that were observed between the different structures were granitic rock, quartz and chalcedony. Map cracking is a common characteristic in the structures studied in Asia. The structures suffered large and long cracks and overall deterioration. Table 10 further draws out the commonalities and differences between the structures in Asia.

Table 10: Comparison of AAR in Asia

Structure	Country	Aggregates used	AAR characteristics	Time from construction to AAR damage	Overall damage
Sewage Works	Hong Kong	Granitic rock	Map cracking	N/A	Large, long cracks on the top surface of concrete wall
Concrete Pavement	Korea	Cryptocrystalline quartz, chalcedony	Cracking	N/A	Deterioration and damage to concrete pavement

2.3.9 AAR in Mainland Europe

Swiss Tunnels – Switzerland

Eight tunnels in Switzerland were studied to verify the existence of AAR damage and to determine the extent of the damage (Leemann, Thalmann, & Studer, 2005). The tunnels were all constructed with concrete and shotcrete, and samples of these were investigated. The tunnels range from 19 to 44 years in age and they range from 0.81 to 15.4 kms in length. The temperatures in the tunnels varied from 12 - 21°C. These tunnels were used for railways, roads or they led water to reservoirs during the rainy seasons (Leemann et al., 2005). The tunnels showed signs of typical crack patterns, pop outs and excessive deposits on the concrete surface, which indicated AAR.

Petrographic examinations revealed that igneous rocks made up the majority of the aggregates with 53-93% by mass. The aggregates also consisted of major sedimentary rock types such as sandstone, limestone, siliceous limestone and slate. Minor amounts of mafic rocks were also found in the aggregates. It was found that the AAR occurs in the concrete tunnels which are between the ages of 19 and 44 years (Leemann et al., 2005). Figure 21 shows the dense cracking pattern seen in Swiss tunnels as a result of ASR damage (Fernandes et al., 2017).

Figure 21: Tunnel showing dense cracking pattern in Switzerland (Fernandes et al., 2017)

Concrete Road and Dam – Austria

ASR was first reported in Austria in 2001 in two major structures. These were: a concrete road surface which was constructed in 1990 and started to show signs of deterioration due to ASR in 1994. The ASR was attributed to the use of reactive carbonate aggregate. The other structure was a dam which was built in 1942 and exhibited signs of severe damage in 1993/1994 and thus had to be replaced with a new structure. At a later stage, two motorways also showed signs of ASR. The first motorway, the A9, was built in 1985 and exhibited map cracking 13 years after construction. The aggregates used in the construction and which were the cause of the ASR were quartzite and gneiss.

Commonalities and differences in the AAR in Mainland Europe

In the structures studied in Mainland Europe, the type of structures affected by AAR included dams, concrete roads and tunnels. Some aggregates that were observed between the different structures were limestone, slate, carbonate aggregate, quartzite and gneiss. It was also seen that for the structures studied in Mainland Europe, the damage due to AAR can occur as early as 4 years from the time of construction. Map cracking, dense cracking and pop outs are very common characteristics in the structures studied in Mainland Europe. Some structures suffered severe deterioration in a short amount of time while others did not suffer substantial damage despite being affected by AAR. Table 11 further draws out the commonalities and differences between the structures in Mainland Europe.

Table 11: Comparison of AAR in Mainland Europe

Structure	Country	Aggregates used	AAR characteristics	Time from construction to AAR damage	Overall damage
Tunnels	Switzerland	Sandstone, limestone, siliceous limestone and slate	Dense cracking, patterns, pop outs and excessive deposits on the concrete surface	Ranging between 19 and 44 years	Although cracking due to AAR was present, no substantial damage was recorded
Concrete road and dam	Austria	Carbonate aggregate, quartzite and gneiss	Map cracking	From 4 years	Severe deterioration

2.3.10 AAR in North Africa

Transmission Tower Bases - Middle East and North Africa

This area has had very few recorded cases of AAR damage. One recent anecdotal case was recently reported on transmission tower bases in Libya (Kay, Poole, & Sims, 2017). A few other reports of possible AAR damage were also made about structures in Bahrain, Yemen and Israel, but no information is available about any of these cases (Kay et al., 2017). There is very limited information regarding AAR in this area.

Summary of AAR in North Africa

Table 12 shows a summary of the details of the structure studied in North Africa.

Table 12: Comparison of AAR in North Africa

Structure	Country	Aggregates used	AAR characteristics	Time from construction to AAR damage	Overall damage
Transmission Tower Bases	Libya, North Africa	Siliceaous rock	ASR gel exudence, random cracking	1 month	Extensive cracking

2.3.11 Discussion and Conclusion

This section highlighted incidences of AAR encountered in concrete structures around the world. The structures were grouped by geographic location with the aim of drawing any commonalities and differences in AAR occurrence in the different parts of the world. A variety of types of structures that were discussed in this section, including: bridges; water retaining structures; power station infrastructure; foundations and bases; sewage works infrastructure; railway infrastructure; tunnels; concrete pavements and car parks. The different areas/regions that the structures were grouped into were: The United Kingdom, Russia, North America, India, Southern and Central Africa, Australia, Nordic Europe, Asia, Mainland Europe and North Africa. One feature that all AAR-affected structures exhibit regardless of the type of structure or its location, is cracking. All AAR-damaged structures suffer from varying degrees of cracking. The cracks range from fine cracks to wide cracks.

Three structures from **The United Kingdom** were studied in this section – a dam, an electricity substation and a car park. All three structures were constructed with concrete containing aggregates with chert, while two of the structures contained limestone. This indicated that these two types of aggregates promote the occurrence of AAR in concrete structures in this region. Other aggregates used in this region which also resulted in AAR were shale, feldspar and quartzite. All three structures experienced slight to severe map cracking while the dam experienced discoloration. The substation experienced reaction rims in the concrete and gel deposits. The dam experienced wall displacements of up to 13mm. Displacement of concrete element was also seen in other water retaining structures from other parts of the world. The amount of time taken for structures to show signs of AAR damage ranged from 7 to 10 years from the time of construction for the structures studied.

One type structure from **Russia** was studied in this section – railway sleepers. This structure was constructed with concrete containing the aggregates flint, sandstone and andesites. This structure experienced cracks with varying width and gel deposits on the concrete. The railway sleepers were severely disintegrated, with crack widths of 1cm. The severe damage of the sleepers can also be attributed to the fact that railway sleepers are thin concrete members, which makes it easier for the entire member to be severely damage once it starts cracking. The amount of time taken for sleepers to show signs of AAR damage was 3 years from the time of construction.

Three structures from **North America (in The United States of America and Canada)** were studied in this section – 2 dams and a bridge deck. All three structures were constructed with concrete containing different types of aggregates, i.e. andesitic aggregate, greywacke and contained limestone respectively. All three structures experienced map cracking. All three structures also experienced significant damage where the structures had a reduction in mechanical properties and a reduction in their service life. The amount of time taken for structures to show signs of AAR damage ranged from 2 to 10 years from the time of construction for the structures studied.

Three structures from **India** were studied in this section – both were dams and their related structures. Both structures were constructed with concrete containing granite aggregates. This indicated that this type of aggregates promote the occurrence of AAR in concrete structures in this region. Other aggregates used in this region which also resulted in AAR were granite, diorites and quartzite. Both structures experienced cracking and gel deposits. Both dams also experienced displacements of concrete elements which resulted in operational issue. Displacement of concrete element was also seen in other water retaining structures from other parts of the world. The amount of time taken for structures to show signs of AAR damage ranged from 25 to 27 years from the time of construction for the structures studied.

Various structures from **Southern and Central Africa (In South Africa, Namibia and Uganda)** were studied in this section including water retaining structures, bridges, power plants, road pavements and underground structures. For many of the structures that were affected by AAR, they were constructed with concrete containing granite and gneiss. Two very common aggregate types also observed were the Witwatersrand Supergroup quartzites and greywacke rocks of the Malmesbury Group. The power plant in Uganda had different aggregates from Southern Africa – schistose aggregates. All the structures experienced cracking. Although structures in this region experienced severe cracking, no substantial damage on their overall performance due to AAR was reported. The amount of time taken for structures to show signs of AAR damage ranged from 3 to 40 years from the time of construction for the structures studied.

One type structure from **Australia** was studied in this section – water retaining structures. This structure was constructed with concrete containing quartz. This structure experienced map cracking. The dam experienced wall displacements and rotations. Displacement of concrete element was also seen in other water retaining structures from other parts of the world.

Various structures from **Nordic Europe (in Denmark and Sweden)** were studied in this section – bridges and other structures. In all the structures in all the areas studied, the structures were constructed with concrete containing aggregates with flint. This indicated that this type of aggregate promotes the occurrence of AAR in concrete structures in this region. In Denmark, the structures experienced map cracking and delamination which were aggravated by de-icing salts, while the structures in Sweden experienced pop-outs. The substation experienced reaction rims in the concrete and gel deposits. The amount of time taken for structures to show signs of AAR damage ranged from 5 to 10 years from the time of construction for the structures studied.

Two structures from **Asia (in Hong Kong and Korea)** were studied in this section – sewage works and concrete pavements. Both structures were constructed with concrete containing different types of aggregates, i.e. granite and quartz respectively. Both structures experienced cracks of varying widths and lengths.

Three structures from **Mainland Europe (Switzerland and Austria)** were studied in this section – a dam, tunnels and a concrete road. All three structures were constructed with concrete containing different types of aggregates, i.e. Sandstone, limestone, siliceous limestone slate, carbonate aggregate, quartzite and gneiss. All three structures experienced map cracking and the tunnels experienced pop outs. No substantial damage was reported in Switzerland, while severe deterioration was reported in Austria. The amount of time taken for structures to show signs of AAR damage ranged from 4 to 44 years from the time of construction for the structures studied.

One type structure from **North Africa** was studied in this section – transmission tower bases. This structure was constructed with concrete containing siliceous rock. This structure experienced random cracks and gel exudence on the concrete. The amount of time taken for sleepers to show signs of AAR damage was only 1 month from the time of construction.

It can be seen that in most part of the world, the types of aggregates that cause AAR are similar in those regions. It can also be seen that the type of damage is similar in the same type of structures, particularly the water retaining structures which exhibit distinct types of damage, which were not visible in other types of structures. These structures were also among those that exhibited AAR damage the earliest regardless of their location in the world. The type of aggregate depends on the geographical location of the AAR-affected structure as most constructions utilize the aggregates that are locally available. One trend that could be seen is that certain aggregates were used in several structures which were affected by AAR. Limestone was used in AAR-affected structures in Canada, Switzerland and The United Kingdom. Sandstone was used in AAR-affected structures in Russia and Switzerland. Granite was used in AAR-affected structures in The United Kingdom, India, Hong Kong, Namibia and South Africa. Quartz was used in AAR-affected structures in The United Kingdom, Korea, India, South Africa, Austria and Australia. Flint was used in AAR-affected structures in Denmark and Russia

3. AAR Avoidance Measures Worldwide

3.1   Introduction

Since AAR is a type of internal chemical damage to concrete, it can be avoided by engineering design and by carefully selecting the concrete construction materials. In order for damaging AAR to occur in concrete, the following conditions need to be met (Sims & Poole, 2017) and (Godart & de Rooij, 2013):

- Reactive silica in the aggregates should be present
- Alkali, which is primarily from Portland cement, should be of a sufficient concentration
- There should be sufficient moisture in the concrete
- Portlandite should be in a sufficient concentration (specifically only for ACR)

To prevent the occurrence of AAR in concrete, one or more of the conditions above should be eliminated, except for the case of ASR in which one or more of the first three conditions should be eliminated. Since this book mainly focuses on ASR, only the first three conditions will be considered, because as explained in Chapter 1, the fourth condition is specific to the occurrence of Alkali-Carbonate Rock Reaction (ACR). The measures discussed in this chapter will reduce the occurrence of ASR by addressing one of these three conditions. This chapter will discuss 7 measures that may be employed to avoid AAR in concrete structures. These measures are:

- The use of non-reactive aggregates
- Limiting alkali content of concrete
- The use of lithium and sodium compounds
- Use of fine lightweight aggregates (FLWAS)
- The use of supplementary cementitious materials
- The use of steel fibres
- Managing environmental exposure

These measures include those that are practically being applied around the world as well as those that have been proven through experiments to reduce or eliminate ASR in concrete. In this chapter, a critical analysis will also be done on all the avoidance measures which are discussed.

3.2   The Use of Non-Reactive Aggregates

Using aggregates which are not deleteriously reactive is a viable option in the avoidance of ASR-induced damage in concrete. This option however depends on the availability of non-AAR reactive aggregates being readily available in the area where construction is taking place. Tests need to be carried out to determine the reactivity of the aggregates, and these tests need to be carried out on a regular basis to ensure that the composition and thus the reactivity of aggregates remains constant within an aggregate quarry. If the non-reactivity of the aggregates can be conformed, then these aggregates can be used without further concerns or need for further ASR preventive measures (Sims & Poole, 2017) . Using non-reactive aggregates is not always a feasible option, due to the following reasons (Sims & Poole, 2017):

- Non-reactive aggregates may not be locally available and high costs would be involved in transporting such aggregates from other locations
- Reactive aggregates may be significantly more affordable and may also have less negative environmental impact than non-reactive aggregates
- Regardless of their true reactivity, all the locally available aggregates may fail the accelerated mortar bar test (ASTM C1260), thus making it impossible to select the correct aggregate

It should be noted that even when using non-reactive aggregates, some construction projects will still require extra measures of caution due to their critical nature. Such projects include critical structures such as those with an extended design life or structures exposed to very aggressive environments such as those exposed to seawater.

3.3 Limiting Alkali Content of Concrete

Extensive studies done on ASR indicated that in order for expansive alkali silica reaction not to occur, the alkali content of the cement needs to be kept below 0.6% Na_2O_{eq}. This has been adopted as the minimum alkali content in cement that is used together with reactive aggregates, and it is stated in ASTM C150 Standard Specification for Portland cement that is used in concrete made with deleteriously reactive aggregates. The value of 0.6% is however an arbitrary value, which is only applicable if the contribution of alkalis from other sources is small and that the cement content is less than 350kg/m^3 (Oberholster, 2009). It has been found that limiting the alkali level of the cement alone does not give an accurate representation of the ASR potential. Alkali content can be influenced by mix water, sea spray as well as chemical admixtures such as accelerators, retarders and workability aids (Oberholster, 2009). Additionally, the reaction of sodium and potassium bearing minerals in the aggregate with $Ca(OH)_2$ released during hydration also produces alkalis.

A more accurate indication is limiting the alkali content of the concrete when reactive aggregate is used in concrete, Na_2O_{eq}/ m^3 concrete. Alkalis may be sourced from the reaction of calcium hydroxide during the hydration of cement with alkali-containing minerals in the aggregates, or it may be sourced from external sources such as mixing water, sea water, sea spray or aggregates which contain salts or from chemical admixtures like sodium lignosulphonate (Oberholster, 2009). Generally though, the alkalinity of the pore solution is determined by the product of the alkali content of the cement and the amount of cement in the concrete, and this is what determines the extent the alkalis will react with the aggregates. It is therefore more accurate to limit the alkali content per m^3 of concrete and not to the cement content (Oberholster, 2009). Due to this, cement-rich concrete mixes will contain higher contents of alkali per m^3 of concrete and will thus be more likely to cause expansion if used in combination with a reactive aggregates.

Figure 22 shows the relationship between the alkali content of varying cement contents and the expansion of the concretes produced with those cement quantities. These figures demonstrate that the alkali content is controlled by the product of the cement alkali level, in other words the alkali content of the concrete (Sims & Poole, 2017).

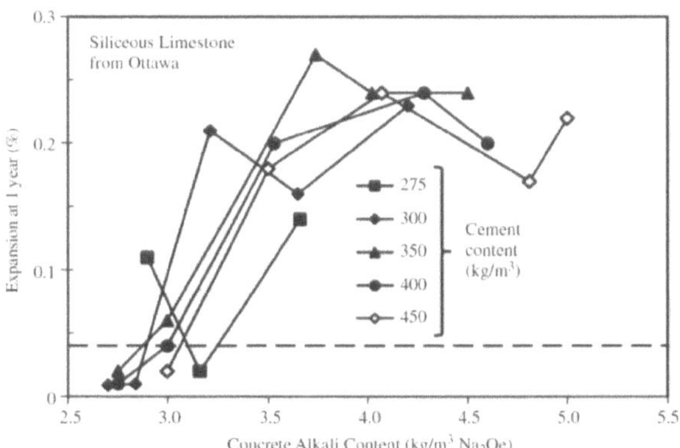

Figure 22: Expansion of concrete prisms as a function of alkali content. The dashed line represents the critical 1 year expansion limit (Sims & Poole, 2017)

Many specifications have employed the maximum concrete alkali content as the measure to control expansion in concretes which contain reactive aggregates. Table 13 shows the specified total alkali limit in concrete in different countries around the world.

Table 13: Alkali limits for controlling expansion from reactive aggregate used internationally (Mackechnie, 2021)

Country	Specified total alkali limit in concrete	Comments and performance guidelines
USA ASTM A23.1&2–2016	$1.8 - 3.0 \ kg/m^3$	Prescriptive and performance approach used. Performance using ASTM C1293 & C1567
Canada CSA A23.1&2–2014	$2.0 - 3.0 \ kg/m^3$	Limits based on environment and risk for each structure
UK BRE Digest 330.2-2004	$3.5 \ kg/m^3$	Alkali limits based on cement alkali limits and aggregate reactivity
RILEM AAR 7.1 – 2016	$2.5 - 3.5 \ kg/m^3$	Limits based on aggregate reactivity, with classification being low, medium or high
Japan JIS A5308-2009	$3.0 \ kg/m^3$	Prescriptive limit based on andesite data. Performance limits from 1.2-3.0 kg/m^3
Australia HB79 – 2015	$2.8 \ kg/m^3$	Risk with reactive aggregate almost always controlled using SCMs [1]
New Zealand CCANZ TR3 – 2012	$2.5 \ kg/m^3$	Prescriptive limit widely used. Performance limits from 1.8-3.0 kg/m^3

1 SCM: Supplementary cementitious material, such as fly ash or slag; also termed 'additions

3.4   The Use of Lithium and Sodium Compounds – New Zealand

The use of lithium and sodium compounds was suggested in concrete to reduce the expansion of concrete that is affected by alkali silica reaction. There was also research work done which confirmed that ASR in concrete can be suppressed by adding sodium and lithium compounds to the concrete. Additionally, research showed that ASR gel which consisted of sodium compounds was less expansive, while that with lithium hydroxide resulted in the decrease in the total amount of ASR gel produced (Prasetia & Torii, 2013).

It was found via research that the most effective lithium compounds in controlling ASR are lithium nitrate ($LiNO_3$) and lithium hydroxide monohydrate ($LiOH.H_2O$). The effectiveness of the lithium compound in reducing ASR depends on the nature of the reactive aggregate in the concrete. It has been seen that in order to prevent ASR, a molar ratio of [Li]/[NA+K] of 0.6 was required (Oberholster, 2009). Incorporating between 0.8 and 1.2% by mass of lithium hydroxide monohydrate with a cement with a high alkali clinker (>1% Na_2O-eq) did not result in the deleterious expansion of concrete due to ASR. Lithium compounds may also be used in combination with ground granulated blastfurnace slag, silica fume or fly ash in prevent ASR.

The most common way in which these compounds are incorporated into concrete is by the addition of chemical admixtures which contain these compounds, during the concrete mixing stage. Construction material manufacturers such as Sika, GCP Applied Technologies and The Euclid Chemical all manufacture lithium-nitrate based chemical admixtures which are added to concrete with the purpose of preventing ASR. The dose of admixture is based on the Na_2O-eq of the cement. For example, the dosage of Sika Control ASR (which is the ASR control admixture produced by Sika) is $4.6kg/m^3$ of Na2O-eq supplied by the cement (Sika, 2014). Sika's admixture to control ASR was use in the Detroit Metropolitan Airport in Michigan, USA (Sika, 2014). When determining the correct dosage of admixture, it is very important to verify the exact quantity by performing mortar tests that assess the expansion of the concrete.

Research was conducted in Japan to investigate the diffusivity of externally supplied lithium and sodium compounds into cementitious material. The purpose of this research was to investigate the possibility of using a more economical and effective compound for the mitigation of ASR in concrete (Prasetia & Torii, 2013). The diffusivity was tested using the diffusion cell test (Prasetia & Torii, 2013). Mortar bar tests were then employed to measure the expansion of test specimens and thus test the mitigation effects of lithium and sodium compounds on ASR in concrete (Prasetia & Torii, 2013).

Figure 23 (Prasetia & Torii, 2013) shows the expansion of the following immersed specimens:
- 0.5 mol/L concentrated Li_2SiO_3 (for low concentration)
- 0.5 mol/L concentrated Na_2SiO_3 (for low concentration)
- 2.5 mol/L concentrated Li_2SiO_3 (for high concentration)
- 2.5 mol/L concentrated Na_2SiO_3 (for high concentration)
- 1 mol/L concentrated $NaNO_3$ (for low concentration)

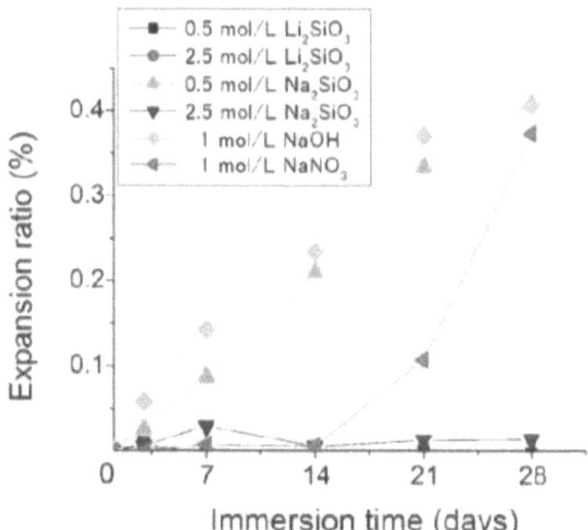

Figure 23: Expansion of concrete prisms as a function of alkali content. The dashed line represents the critical 1 year expansion limit (Sims & Poole, 2017)

The research yielded results which are summarized as follows (Prasetia & Torii, 2013): Specimens which were immersed in Li_2SiO_3 exhibited no expansion, regardless of the concentrations. This may be taken to mean that no ASR took place in these specimens.

- Specimens that were immersed in low concentration (0.5 mol/L) Na_2SiO_3 solution exceeded 0.2% expansion at 14 days and exceeded 0.4% expansion at 28 days

- Specimens that were immersed in low concentration (1 mol/L) NaNO3 had no expansion up to 14 days but had an expansion exceeded 0.3% at 28 days

- Specimens that were immersed in high concentration (2.5 mol/L) Na_2SiO_3 solution had a very small early expansion, which disappeared. Unlike with the low concentration Na_2SiO_3, which was in a liquid state at 80 °C, the high concentration Na_2SiO_3 solution had a high viscosity and was in a paste-like state until 14 days.

- A likely explanation for the Na_2SiO_3 solution behaviour is that it was difficult for the highly concentrated, water-glass solution to penetrate the test specimens while in the presence of the highly reactive flint aggregates. The hydrolysis reaction between the sodium compound and the water produces NaOH, which increases the alkalinity of the pore solution. It also increases the OH- in the pore solution, starting the ASR. Increasing the concentration of Na_2SiO_3 from an external source however, triggers the reaction between Na_2SiO_3 and Ca $(OH)_2$ and it surpasses the pace of the alkali silica reaction. An overwhelming amount of C-S-H gel is produced and engulfs the ASR gel produced previously, thus suppressing the expansion. This activity is different from that of the lithium compounds due to the fact that highly reactive lithium stops ASR from occurring in the first place.

It can be concluded from the experiments conducted that the diffusion coefficients of Na+ ions were 10 times greater than those of Li+, which indicates that lithium compounds are better absorbed into the hydration compounds and they react better with them, compared to Na+ ions. Additionally, no expansion of the mortar bar is observed regardless of the Li_2SiO_3, which shows that there was no ASR took place and therefore Li_2SiO_3 can be used to mitigate expansion due to ASR in concrete (Prasetia & Torii, 2013). Na_2SiO_3 may also be used to mitigate ASR-induced expansion in concrete when applied at high concentration. These measures may be used as cost effective ASR avoidance/mitigation measures in concrete.

The details of this research can be found in Appendix 1.

3.5 Use of fine lightweight aggregates (FLWAs) – United States of America

Fine lightweight aggregates (FLWAs) such as expanded slate, shale and clay have been found to mitigate alkali silica reaction (ASR) in concrete. Expanded clay was found to be the most effective one in reducing ASR. Expanded shale and clay reduce the alkalinity of the concrete pore solution and they also increase the aluminium content of the pore solution (Li, Thomas, & Ideker, 2018). FLWAs are mainly manufactured by heating the raw materials to very high temperature of over 1100°. The heated raw materials liquefy and air bubbles are released from the aggregated, resulting in the expansion of the aggregate (Li et al., 2018). The mechanisms by which FLWAs reduce expansion due to ASR is a subject that is not fully understood and researchers have not been able to reach an agreement on it.

Research was conducted to determine whether fine lightweight aggregates (FLWAs) can be used to mitigate alkali silica reaction in concrete. In the research, three FLWAs were investigated for their mitigation impacts, namely: expanded slate, shale and clay

The accelerated mortar bar test (AMBT) in accordance with ASTM C1260 was employed to test the reactivity of the aggregates, while the concrete prism test (CPT) was also employed to evaluate the aggregate reactivity over a one year period or mitigation efficacy over a two-year period. The results for the AMBT are shown in Figure 22. The FLWAs were prepared either pre-wetted or oven dried and they were used to replace FA1 at 25%, 50% and 100% by volume. Figure 24 shows the expansions of the two pre-treatment groups (pre-wetted and oven-dried) of samples after they were exposed to 1.0 N NaOH for 14 days. The results are summarized as follows:
- The reference sample with the highly reactive fine aggregate FA1 had an expansion of 0.66%, which significantly exceeded 0.1% which is the expansion limit.
- With the exception of the 25% pre-wetted expanded slate, all expansions were reduced with an increase in FLWA replacement in both the pre-wetted and oven dried groups.
- In both groups, the mix with 100% expanded shale was below the expansion limit of 0.1%, indicating that expanded shale was non-reactive

- Expanded clay and expanded shale limited expansion to a better extent that expanded slate
- All mixtures with pre-wetted FLWAs with 100% replacement of FLWAs showed behaviour which was not deleteriously reactive.
- For the oven-dried mixtures, the 25% expanded slate had an expansion of 0.55%, while the 25% expanded slate of the pre-wetted mixtures had an expansion of 0.77%
- The AMBT results show that the mixtures containing the pre-wetted FLWAs had a higher expansion than those containing oven-dried FLWAs at the same quantity of replacement.

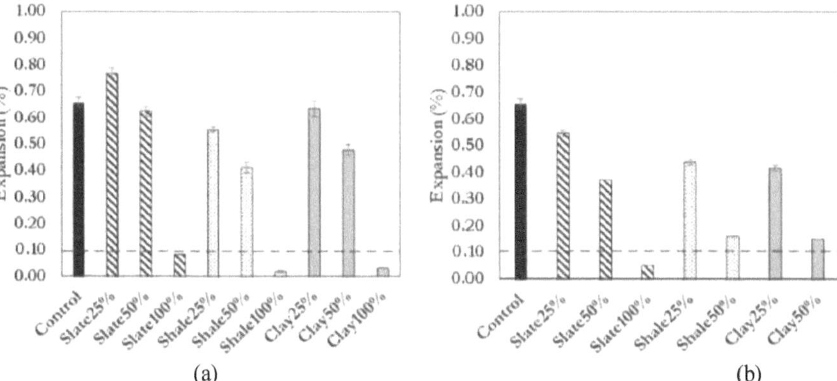

(a)   (b)

Figure 24: AMBT expansion results after 14 days. (a) Mixtures with pre-wetted FLWAs and (b) Mixtures with oven-dried FLWAs (Li et al., 2018)

The results of the CPT for two years are shown in Figure 25. In these tests, all the FLWAs were pre-wetted so that the moisture content was above absorption capacity. The results of the CPT are summarised as follows:
- In the BN mixtures, the reference showed an expansion of 0.66%
- The 25% expanded slate showed a minimal impact on expansion
- From all three FLWAs, expanded clay was most effective in reducing expansion due to ASR
- Mixtures in which FLWAs replaced highly reactive fine aggregates showed a reduction in expansion in both the AMBT and the CPT.

From the results of this investigation, the following is concluded about the impact of FLWAs on mitigating expansion due to ASR:
- Expanded slate, shale and clay all effectively reduce the concrete expansion due to ASR
- An increase in the replacement level of FLWAs results in less expansion in concrete.

Further details of this research can be found in Appendix 2.

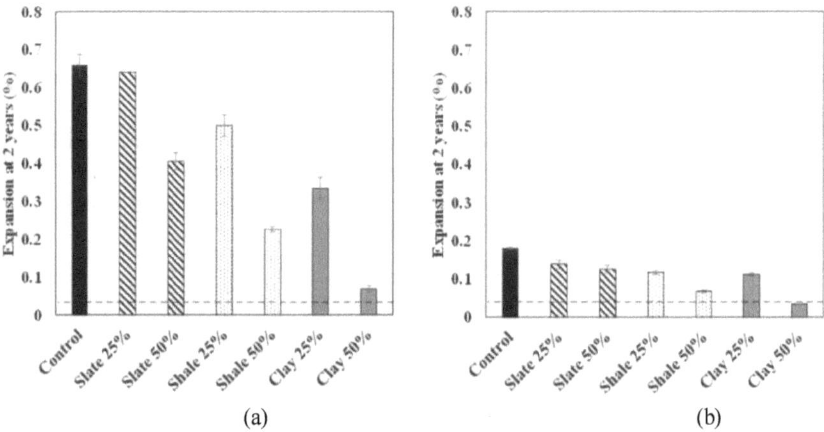

Figure 25: Expansion results of CPT at 2 years. (a) BN mixtures and (b) SP mixtures (Li et al., 2018)

3.6   The Use of Supplementary Cementitious Materials

Supplementary cementitious materials (SCMs) may be used for the purpose of minimizing the risk of AAR in concrete. In South Africa in particular, three main SCMs are used for this purpose, namely fly ash, ground granulated blastfurnace slag and condensed silica fume (Oberholster, 2009). Expansion due to AAR has been successfully reduced by blending a SCM with a high alkali cement when used in combination with a reactive aggregate. The reason for this reduction is the fact that the blending results in the dilution of the cement and therefore a reduction in the content of active alkalis in the concrete. In order to prevent the deleterious expansion of concrete when using alkali-reactive aggregates together with high-alkali cement, the cementitious material should contain a minimum amount of SCM by mass (of the total mass of the cement) of the following (Oberholster, 2009):

- 15% condensed silica fume, or
- 40% ground granulated blastfurnace slag, or
- 20% fly ash

It is further recommended that if the Na_2O-eq of the cement is greater than 1%, then the content of ground granulated blastfurnace slag should be increased by 10% (Oberholster, 2009). In addition to the SCM discussed above, other SCMs have also been considered and used worldwide, namely metakaolin and amorphous rice husk. The SCM replacement level depends on certain factors, including (Thomas, Hooton, & Folliard, 2017):

- The nature of the SCM: a higher content of SCM is required when the content of silica decreases or when the contents of alkali and calcium increase
- The nature of the reactive aggregate used: when the aggregate is more reactive, generally a higher content of SCM is required

- The alkalis in the concrete (from the different possible sources of alkali in the concrete such as the cement and other sources): if there are more alkalis in the concrete, the content of SCM required increases
- The conditions of exposure of the concrete: if the concrete is exposed to external alkali sources, the SCM content required is higher.

All these SCMs will be discussed in more detail below.

3.6.1   Fly Ash and Metakaolin – Portugal

The use of fly ash is one of the most widely used measure to minimise alkali silica reaction (ASR). A research was conducted to study the influence of supplementary cementitious materials such as fly ash and metakaolin on inhibiting the occurrence of ASR in concrete (Silva, Riberio, Jalali, & Divet, 2006). From this research, the results in Figure 26 were obtained and the following conclusions were drawn:

- The expansion of the control increases over time due to the reactive nature of the aggregate used in the mix
- 20% replacement with FA was effective in inhibiting the expansion due to ASR
- 20% replacement with MK had an even more significant effect on inhibiting the expansion due to ASR than FA

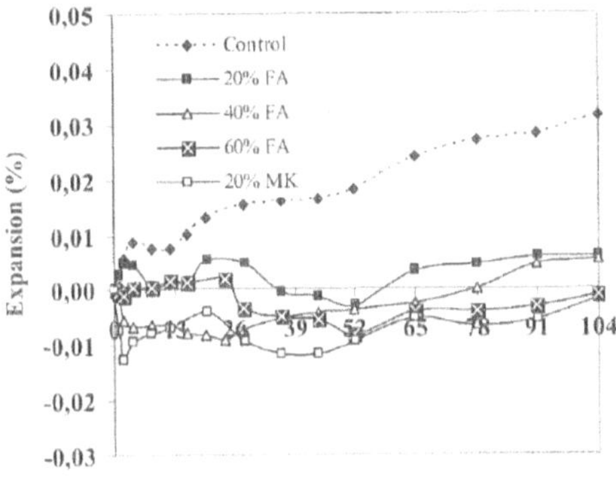

Figure 26: Expansion of concretes in accordance with the RILEM AAR-3 method (Silva et al., 2006)

Further details of this research are shown in Appendix 3.

3.6.2   Ground Granulated Blastfurnace Slag (GGBS) - Canada

A 20-year field evaluation was conducted to evaluate measures to prevent ASR in concrete. Alkali silica reactive aggregates were used to make 6 concrete mixes with different cements and different supplementary cementitious materials. The concrete samples made were placed outside, exposed to environmental conditions with no protective measures and then monitored over a period of 20 years (Hooton etal., 2013). The high alkali cement that had a cement replacement of 25% GGBS showed only minor cracking. A cement replacement of 50% GGBS showed no sign of ASR cracking in all the 20 years during which the monitoring was done on the specimens (Hooton et al., 2013). Figure 27 shows the expansion of these laboratory concrete samples which were stored at a site in Kingston, Canada. These sample with GGBS (Slag) at 25% and GGBS at 50% in comparison to control samples as well as samples with other supplementary cementitious materials.

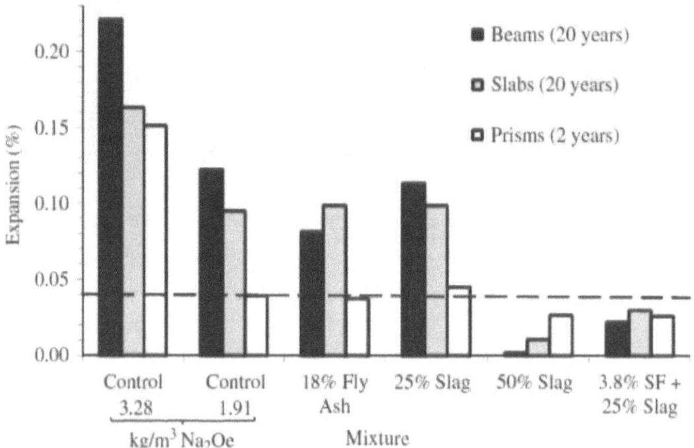

Figure 27: Expansion of concrete prisms in the laboratory and concrete blocks and slabs on the Kingston exposure site (Thomas et al., 2017)

3.6.3   Amorphous Rice Husk-Ash – Brazil

Brazil is one of the world's largest producers of rice, where about 11 million tons of rice are produced every year (Hasparyk, et al., 2004). When rice is processed, its husk is removed. Rice husk does not degrade easily and storing it presents environmental threats. Every ton of husked rice produces 200 kg of husk, which translates to 40 kg of husk ash through combustion. Experiments were carried out to investigate the impact of rice husk ash (RHA) on concrete properties, among them the impact on Alkali Aggregate Reaction (AAR). The tests were carried out by partially substituting the cement with varying quantities of RHA, which were: 8%, 10%, 12%, 16% and 20% (Hasparyk et al., 2004). The RHA was used in two states, namely the natural state and the ground state.

The AAR expansion reduction potential was determined relative to the reference sample and presented in percentage. The results for both types of ash are shown in Figure 28 and Figure 29 at tests ages of 16 and 30 days.

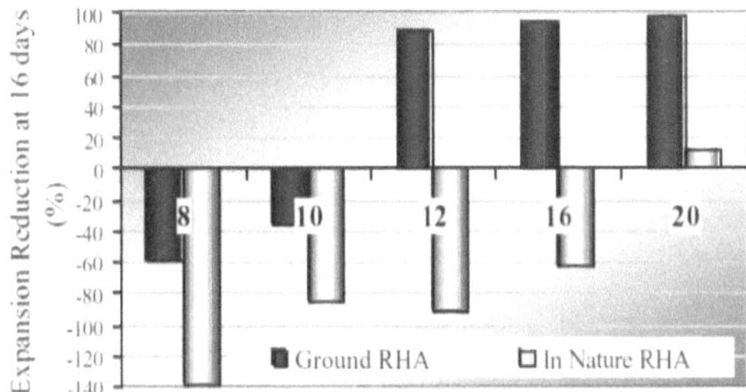

Figure 28: Reduction in expansions of the reference sample, with respect to different contents of RHA at 16 days of testing (Hasparyk et al., 2004)

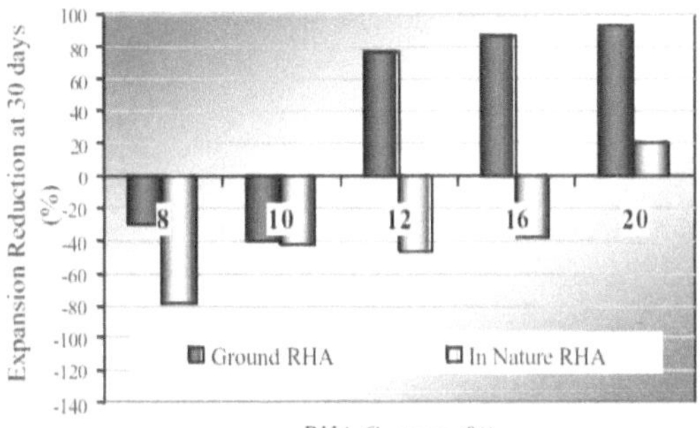

Figure 29: Reduction in expansions of the reference sample, with respect to different contents of RHA at 30 days of testing (Hasparyk et al., 2004)

The following is summarised from the results (Hasparyk et al., 2004):

- In the specimens with ground RHA, the ones with 12%, 16% and 20% contents showed a significantly reduced expansions below 0.1% which is the limit as prescribed by ASTM C1260. This applies to both 16 days and 30 days. Increasing the content of RHA increased the effectiveness of the expansion reduction.
- In the specimens with natural RHA, all the contents (with the exception of 20%), the presence of RHA did not reduce expansions at 16 days. Only a content of 20% RHA showed a reduction in expansion, and even that expansions were above the prescribed limits, meaning that reactive behaviour was still present.

It can be seen from the results of this experiment that ground RHA, when used as a partial replacement of cement can successfully reduce expansion due to AAR. This effect can be seen from a replacement value of 12% and increases with increasing RHA content.

Further details of this research are shown in Appendix 4.

3.7   The Use of Steel Fibres - Brazil

It has been found through research that using steel microfibres in concrete in volumes ranging from 1 to 7% of the concrete significantly reduces the ASR expansion and cracking in concrete (de Carvalho et al., 2010). Microfibres were found to be more efficient than conventional fibres due to their small lengths which allows them to be placed close to the interface of the reactive aggregates where they can influence early gel formation (de Carvalho et al., 2010). Through experimentation, a microfiber volume of 0.5% was observed to have reduced the expansion by 25%. In other tests, a microfibre content of 7% was observed to provide beneficial expansion effects which result from the fact that the ASR gel produced was mechanically confined. The extent of concrete cracking was therefore limited by the fibres (de Carvalho et al., 2010).

An experiment was conducted by the Federal University of Rio de Janeiro in Brazil in conjunction with the Brazilian hydropower industry, to test the impact of steel fibres on mortar specimens which were subjected to Alkali Aggregate Reaction (AAR) (de Carvalho et al., 2010). Two types of steel fibres were used in the experiment, and the fibre contents used were 1% and 2%. The two types of steel fibres used were of dimensions:

- 0.16 mm diameter and 6mm length, and
- 0.2 mm diameter and 13mm length.

The experiment showed that expansion due to AAR may be significantly reduced by up to 61% by adding steel fibres to the concrete in contents of 1% or 2% (de Carvalho et al., 2010). The results can be seen in Figure 30.

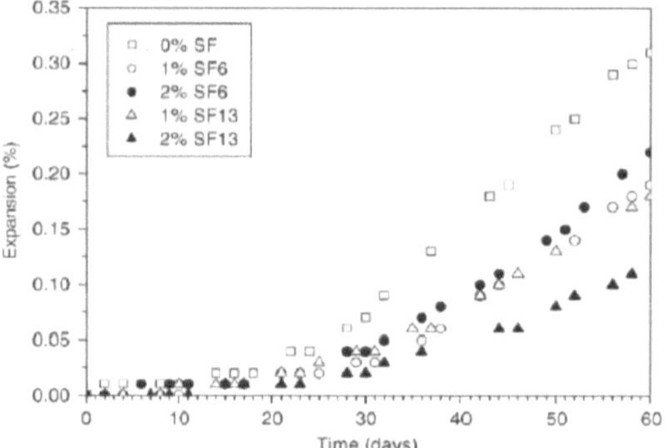

Figure 30: Expansion due to AAR (de Carvalho et al., 2010)

Further details of this research are shown in Appendix 5.

3.8   Managing Environmental Exposure

3.8.1   Moisture

Continuous exposure of concrete elements to moisture makes them more vulnerable to ASR, as moisture is one of the conditions required for ASR to take place. If concrete is protected and kept dry, then any aggregate that complies with standard aggregate requirements (such as SANS 1083: Aggregates from Natural Sources – Aggregates for Concrete (SABS, 2013)) may be used, regardless of reactivity or the alkali content in the concrete mix (Oberholster, 2009). Concrete elements constructed with alkali-reactive aggregates and high alkali cement can be protected from AAR expansion by limiting moisture through improved drainage, cladding or through treatment with a suitable hydrophobic material such as silane or siloxane (Oberholster, 2009).

3.8.2   Exposure to sea spray

When reactive aggregate is used in a high alkali concrete mix, exposure of the concrete to sea spray increases concrete expansion, particularly in the first 90 mm of the concrete. This is due to the fact that the sea spray increases the sodium content of the concrete (Oberholster, 2009). When reactive aggregate is used in a marine environment, the alkali contribution from the sea spray need to be considered in the total alkali calculation of the concrete in the long term. In the presence of sea spray, the average alkali content may increase by 1.5kg Na_2O-eq/m^3 in 10 years (Oberholster, 2009).

3.9   Critical Analysis of AAR Avoidance Measures Discussed

This chapter discussed 7 measures that may be employed to avoid AAR in concrete structures. These measures are:
-   The use of non-reactive aggregates
-   Limiting alkali content of concrete
-   The use of lithium and sodium compounds
-   Use of fine lightweight aggregates (FLWAS)
-   The use of supplementary cementitious materials
-   The use of steel fibres
-   Managing environmental exposure

An analysis will now be done on each avoidance measure discussed in this chapter.

Using non-reactive aggregates is one of the most direct measures to avoid AAR in concrete as it directly removes one of the conditions required for AAR to occur, provided that aggregate sources are thoroughly, competently and regularly tested to confirm non-deleterious reactivity of the aggregates. If it is possible to employ this avoidance measure, then there is usually no need for further precautionary measures to be adopted. The challenge with this measure however is the fact that it is only feasible when the non-reactive aggregate is available in or close to the construction area. Otherwise high transportation costs may be incurred for transporting such aggregates to the desired location. It is for this reason that this option is not a viable option for every construction project and there are areas in the world where those constructing are forced to use deleteriously reactive aggregates in their concrete as these are the only aggregates in their vicinity. In such a case, it is then necessary to employ one of the other AAR avoidance measures.

Limiting the alkali content of the concrete is also a measure to avoid AAR. The main source of alkalis in the concrete is the cement, although alkalis can also be supplied by supplementary cementitious materials, some aggregates, deicing salts, some chemical admixtures and seawater. With this method, many variables need to be controlled in order to ensure that the overall alkali content in the concrete is kept within allowable limits. The alkali content in cement for example is not easy to reduce as it depends on the composition of the raw materials used in the manufacturing process of the cement. The alkali content in aggregates also depends on the aggregate source in the area of construction. Similarly, supplementary cementitious materials and admixtures are supplied as they are by their manufacturers and it would be challenging to alter their composition. Therefore, this option is not viable in all instances, however in instances where the overall alkali content may be kept within allowable limits, this is a very good measure to implement to avoid AAR.

Extensive research has been done on **using lithium and sodium compounds** in preventing ASR in concrete structures. These researches have proven that the use of these compounds definitely reduces the risk of AAR. Practically, lithium compounds have been successfully used in concrete in the form of chemical admixtures which contain the compounds. Most admixtures are lithium nitrate based liquid admixtures that are added to the concrete in specified quantities. Caution should be practiced when using these admixtures because with most of them, the amount of admixture recommended to be added depends of the alkali content (equivalent $Na2O_{eq}$) of the cement. This may not be an adequate way of determining the admixture dosage because as mentioned earlier in this book, the total alkali content of the concrete is provided by various sources including the cement, supplementary cementitious materials, admixtures and deicing salts. These should be taken into account when determining the admixture content required for preventing AAR.

The use of fine lightweight aggregates (FLWAs) has been proven via experimentation to be a feasible strategy to prevent expansion due to AAR. There have not been many reports worldwide about the practical applications of FLWAs. This may be due to the fact that the production of FLWAs is very energy-intensive and may therefore not be economical in practice. The research also showed that in concrete with a high content of cement and/or reactive aggregates, this method may require to be combined with other AAR mitigating strategies in order to be effective. This method still requires further research to determine and define the practicality of its use in construction.

Using supplementary cementitious materials (SCMs) is the most straightforward, efficient and most commonly used method of AAR avoidance when reactive aggregates and/or high alkali cement are used in concrete. Using SCMs reduces the total alkali content of the cement by replacing some of the cement in the concrete and therefore 'diluting' it. Due to its wide usage, there are also extensively published guidelines on sufficient replacement levels as well as considerations that need to be made in order to determine the sufficient cement replacement depending on the SCM being used. This makes them very convenient and easy to apply in practice. SCMs such as ground granulated blastfurnace slag, condensed silica fume, fly ash and metakaolin are widely available all around the world. When considering the use of SCMs, it is important to employ ASR tests to determine whether the proposed concrete mix design meets the ASR mitigation requirements. Rice husk ash is also a SCM which was considered in this book, and it has been found to perform well to prevent ASR. Using rice husk ash in concrete is a feasible preventive measure particularly in rice-producing countries such as China, Brazil and other South American countries. In those countries, large amounts of rice husk is produced as a waste product of rice production, and rice husk ash is then produced as a byproduct of burning rice husk, which is done to reduce the volume of rice husk. Using rice husk ash reduces the burden on the environment.

The use of steel fibres as a method for preventing AAR was proven to be effective through experiments. However, the practicality of this method still requires further studies because the process by which the prevention occurs is not well understood and there are very few researches dedicated to this kind of research. This method of AAR prevention is not widely used around the world and in construction. Other AAR prevention methods are preferred and used instead.

Managing environmental exposure of concrete structures and elements is an avoidance method which is effective if it is possible to control those environmental conditions, and this may not always be the case. If a structure is located in an area where the moisture and sea spray is in contact with the concrete (such as structures close to or in the ocean), then controlling their exposure may be very difficult and the concrete mix design will need to be done taking those conditions into account. However, in circumstances where moisture can be removed from structure by means such as drainage, these may be employed and they will assist in successfully preventing AAR.

3.10 Discussion and Conclusion

The 7 avoidance measures discussed in this chapter all have different aspects about them which make their application practical in different projects. The key to the successful implantation of any of these methods is to test the method for a particular project and particular concrete mix design before practically implementing it. The various testing methods that can be implemented for this are discussed in the next chapter. Table 14 is a summary of each avoidance measure discussed. The summary describes the strengths of each method in terms of ease of application, cost and practicality. The summary also describes the challenges that may be faced when considering to implement each avoidance measure.

Table 14: Summary of Strengths and Challenges of AAR Avoidance Measures

Avoidance Measure	Strengths	Challenges
The use of non-reactive aggregates	When implemented, it completely eliminates the possibility of AAR	Only feasible when non-reactive aggregates are locally available, otherwise high transport costs may be incurred.
Limiting alkali content of concrete	If the alkali content of the concrete can be kept below the recommended limits, then this method alone is sufficient without having to employ other methods.	Alkalis can be supplied to the concrete pore solution in a variety of ways such as cement, supplementary cementitious materials, some aggregates, deicing salts, chemical admixtures and seawater. All these factors need to be considered to control the total alkali content in the concrete.
The use of lithium and sodium compounds	Allows for the use of available aggregates and cement without having to alter them or source alternatives.	Depends entirely on the correct dosage of the compound containing lithium or sodium compounds (i.e. admixtures). The supplier's recommended dosage may not be sufficient because it is usually based on the alkali content of the cement, whereas the alkali content of the concrete is what is more accurate.
Use of fine lightweight aggregates (FLWAs)	Experiments have shown a lot of practical potential of this method.	The production of FLWAs is energy-intensive and therefore costly. Additionally, not enough research has been conducted about this method, and from the existing research, there is a need to combine this measure with other avoidance measures.
The use of supplementary	The most commonly used and most practical avoidance measure. It is also more cost-	Using some SCMs such as rice husk ash is only viable in

cementitious materials (SCMs)	effective compared to other methods. SCMs are widely available all over the world. There are good guidelines with regards to the cement replacement with different SCMs, making it easy to implement this measure. Using rice husk ash has added environmental benefits since rice husk is a waste material which would otherwise need to be burnt. Fly ash and GGBS are waste materials and would otherwise need to be disposed of in waste dump sites, causing an environmental burden.	countries which are large producers of rice.
The use of steel fibres	Has been proven to be effective through experimentation.	This method has not had much practical applications as further research still need to be conducted
Managing environmental exposure	If it is possible to limit or control environmental exposure, then no other measures need to be implemented.	Only effective if it is possible to continuously and consistently manage and control environmental conditions such as moisture and sea spray.

4. AAR Testing Methods Worldwide

4.1 Introduction

Various testing methods are employed all over the world to assess AAR. In this chapter, these tests are reviewed. The testing methods discussed in this chapter include:
- Tests used widely around the world,
- Tests that have been proven and suggested via experiment,
- Tests that were practically used in some of the ASR incidences in Chapter 2 of this book.

The following testing methods will be discussed in this chapter:
- Integrated Schemes According to RILEM and North American Guidelines, comprising three assessment stages, namely:
    o Petrographic examination;
    o Accelerated mortar-bar test for screening and
    o Concrete prism tests
- Concrete Performance Tests
- AFNOR P 18-588 Microbar Test used in Switzerland
- Concrete Imaging performed
- Stiffness Damage Test (SDT) and Damage Rating Index (DRI)
- Multi-Physics Approach (Non-Linear Acoustic Measurements and Microwave Materials Characterisation Measurements)

The tests mentioned above include tests performed to assess whether certain aggregates are susceptible to AAR; tests to assess the performance of specific concrete mixes and thus determine if they are susceptible to AAR, and also tests performed to assess the occurrence and extent of AAR in existing concrete structures.

The testing methods will be discussed in detail, a critical analysis will be performed on them and their strengths and limitations will be highlighted. Any similarities and differences in the methods, will also be highlighted.

4.2 Integrated Schemes According to RILEM and North American Guidelines

Due to the fact that ASR is a slow reaction and may take many years for the damage it causes to be visible, the test methods that have been developed to identify susceptible aggregates over years always attempted to accelerate the results and therefore significantly reduce the testing time. This was done by intensifying one or more of the parameters that contribute to the rate of ASR damage, such as increasing the temperature, increasing the availability of water, increasing the alkalinity around the aggregate or increasing the aggregate fineness (Nixon & Fournier, 2017). The downside to doing this however, was that as the reaction was accelerated, the resemblance to the actual conditions experienced by the aggregate decreased, thereby also decreasing the reliability of these testing methods. In some cases, the tests would identify certain aggregates as reactive while they were in fact safe to use in practice, whereas in other cases, these significantly accelerated tests identify aggregates as unreactive while they actually cause damage when used in construction. This normally happens with aggregates that react very slowly and the damage only becomes visible in the concrete structures after 20 to 50 years (Nixon et al., 2017).

Over the years, as the danger of AAR became more recognised, various research and scientific bodies all over the world attempted to develop accelerated tests which would be able to identify aggregates which were susceptible to AAR and therefore a threat to concrete structures. In the 1980s, South Africa was the first country to develop the accelerated mortar bar test (AMBT), which formed the basis of other AAR tests worldwide and later became standardised with various standards bodies (Hooton & Rogers, 2003). This is elaborated on in Section 4.2.2. Some of the other early standards were developed in the USA and these were the ASTM C227 Mortar-Bar Test and the ASTM C289 Chemical Method. Both of these standards have now been withdrawn and are no longer in use. ASTM C289 was withdrawn in January 2016 (ASTM, 2016). ASTM C227 was withdrawn in October 2018 due to its limited use in industry (ASTM, 2019). These early testing methods were effective in identifying some of the aggregates which caused ASR damage, such as siliceous aggregates containing fast reacting opal or volcanic glass. However, wider use of the methods revealed that these methods could not be applied on a universal scale, as they were not effective in identifying slowly reacting aggregates such as the highly reactive greywacke/argillite aggregate which was responsible for the ASR damage in the famous Mactaquac dam in New Brunswick, Canada (Nixon & Fournier, 2017). The details of the ASR damage on this dam are highlighted in Chapter 2 of this book.

Consequently, many countries all over the world started to develop testing methods tailored to their specific aggregate types, such as the opaline sandstones in north Germany. This led to the development of various testing methods such as ultra-fast autoclave methods and other methods which utilised large concrete specimens. The chemical shrinkage method was also used in Denmark (P. Nixon & Fournier, 2017). These methods yielded reliable results for the areas that they were tailored to, but they did not give reliable results when applied to other areas that had different geologies. It was because of this that the first RILEM Technical Committee on AAR (RILEM TC 106, Accelerated tests for aggregate reactivity) was formed (P. Nixon & Fournier, 2017). At the time that this was happening, North America was also experiencing the same problems with regards to not being able to apply the tests across the board, which then resulted in the ASTM and the Canadian Standards reviewing their testing methods as well. Eventually, the three bodies – ASTM (which is used by many countries around the world), CSA (which is specifically used in Canada) and RILEM (which is used by many countries around the world) - arrived at a consensus and recommended an assessment methodology that comprises three stages, which are (Nixon & Fournier, 2017):
1. Petrographic examination
2. Accelerated mortar-bar test for screening

3. Concrete specimens for long term testing

Figure 31 shows the integrated scheme according to the RILEM methods, while Figure 32 shows the integrated method according to the North American guidelines (Nixon & Fournier, 2017).

The tests conducted under the integrated scheme according to RILEM methods shown in Figure 31 are:
- AAR-0: Overall Assessment Guide on Using RILEM Methods
- AAR-1.1: Detection of Potential Alkali-Reactivity - RILEM Petrographic Examination Method
- AAR-1.2: Petrographic Atlas
- AAR-2: Detection of Potential Alkali-Reactivity - Accelerated Mortar-bar Test Method for Aggregates
- AAR-3: Detection of Potential Alkali-Reactivity - 38 °C Test Method for Aggregate Combinations using Concrete Prisms
- AAR-4.1: Detection of Potential Alkali-Reactivity - 60 °C Test Method for Aggregate Combinations using Concrete Prisms
- AAR-5: Rapid Preliminary Screening Test for Carbonate Aggregates

The principle by which the integrated scheme according to RILEM methods operates is briefly discussed (Nixon & Sims, 2016):
1. The first step in assessing an aggregate combination for AAR potential is petrographic examinations of the aggregates to determine the composition of those aggregates and determine the types and the concentrations of the different reactive components in those aggregates.
2. Once this assessment is completed, the aggregates are then placed in one of the three following categories:
    - Class I for aggregates which are 'very unlikely to be alkali-reactive',
    - Class II for aggregates which are 'potentially alkali-reactive or alkali reactivity uncertain'
    - Class III for aggregates which are 'very likely to be alkali-reactive.
   Class II is normally the classification given to new aggregate sources, after which further tests need to be conducted. In the case of existing aggregate sources where there is past experience of their use in the location under consideration, Classes I and III are the classifications that are normally used.
3. When the petrographic test indicates that the aggregate is either classified in Class II or Class III, the following step is deciding on the appropriate tests to conduct further. The aggregates which are mainly siliceous, or those that are carbonates with a potentially reactive silica content are further classified into Class II-S or Class III-S, and they may then be tested in accordance with RILEM expansion tests (AAR-2, AAR-3 or AAR-4.1).
4. The aggregates which are mainly carbonate, or those that contain reactive types of carbonate are further classified into Class II-C or Class III-C, and they are subjected to specialised procedures for aggregates containing carbonate materials.

5. Aggregates with a composition of both silica and carbonate are designated Class II-SC or Class III-SC and are tested in accordance with RILEM expansion tests (AAR-5).

6. The silica content that can cause very damaging reaction depends on the reactivity of the silica. A small content of highly reactive silica in the aggregate will cause the most damage, while an aggregate containing a high content of highly reactive silica can cause little damage. Mixing aggregate with highly reactive silica with non-reactive aggregate will result in behavior which may either be very damaging, not damaging at all or anything in-between, depending on the different silica proportions in the concrete mix. This is referred to as the pessimum effect. Due to this, it is important that the entire aggregate combination (coarse and fine aggregate) is assessed. AAR-3 and AAR-4.1 can be used for the assessment of combined aggregate.

7. The AAR-3 concrete prism test method was previously considered the reference test due to its use in various forms. However, this test required a long duration of up to 12 months or more. AAR-4.1 required up to 4 months in order to obtain reliable results.

8. As a result, the mortar bar test (AAR-2) and the concrete bar test (AAR-5) were developed in order to obtain results earlier.

9. It is considered unreliable to solely use results from the petrographic assessment where accelerated screening tests are used. When these tests are performed, it is advised that they be confirmed with one of the concrete prism tests.

10. Practical experience has also shown that the accelerated mortar bar test (AAR-2) may not be reliable for Class II-S aggregates that contain porous flint.

The three stages of assessment (Petrographic examination, Accelerated mortar-bar test for screening and Concrete specimens for long term testing) in the integrated schemes will now be discussed (Nixon & Fournier, 2017).

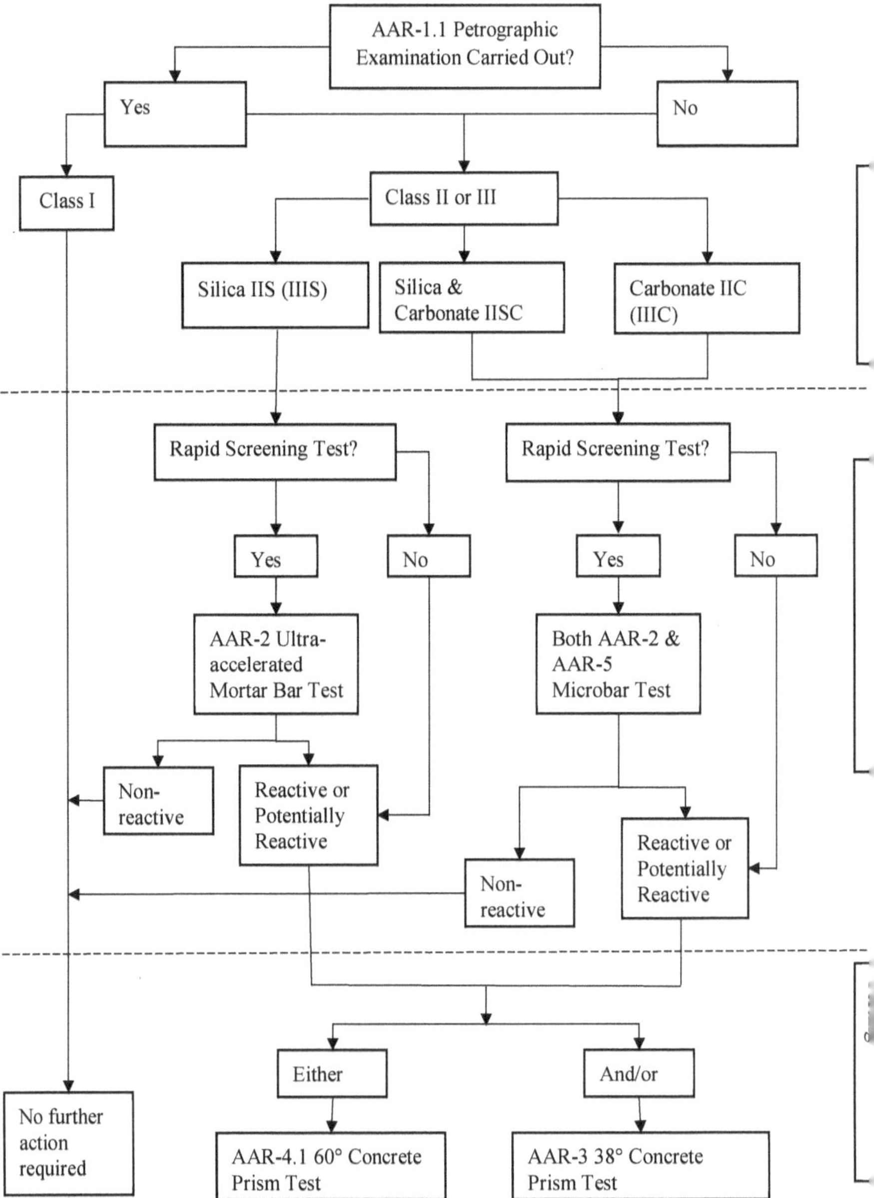

Figure 31: Integrated aggregate assessment scheme for the assessment of alkali-reactivity potential of aggregates (RILEM AAR-0) (Nixon & Fournier, 2017)

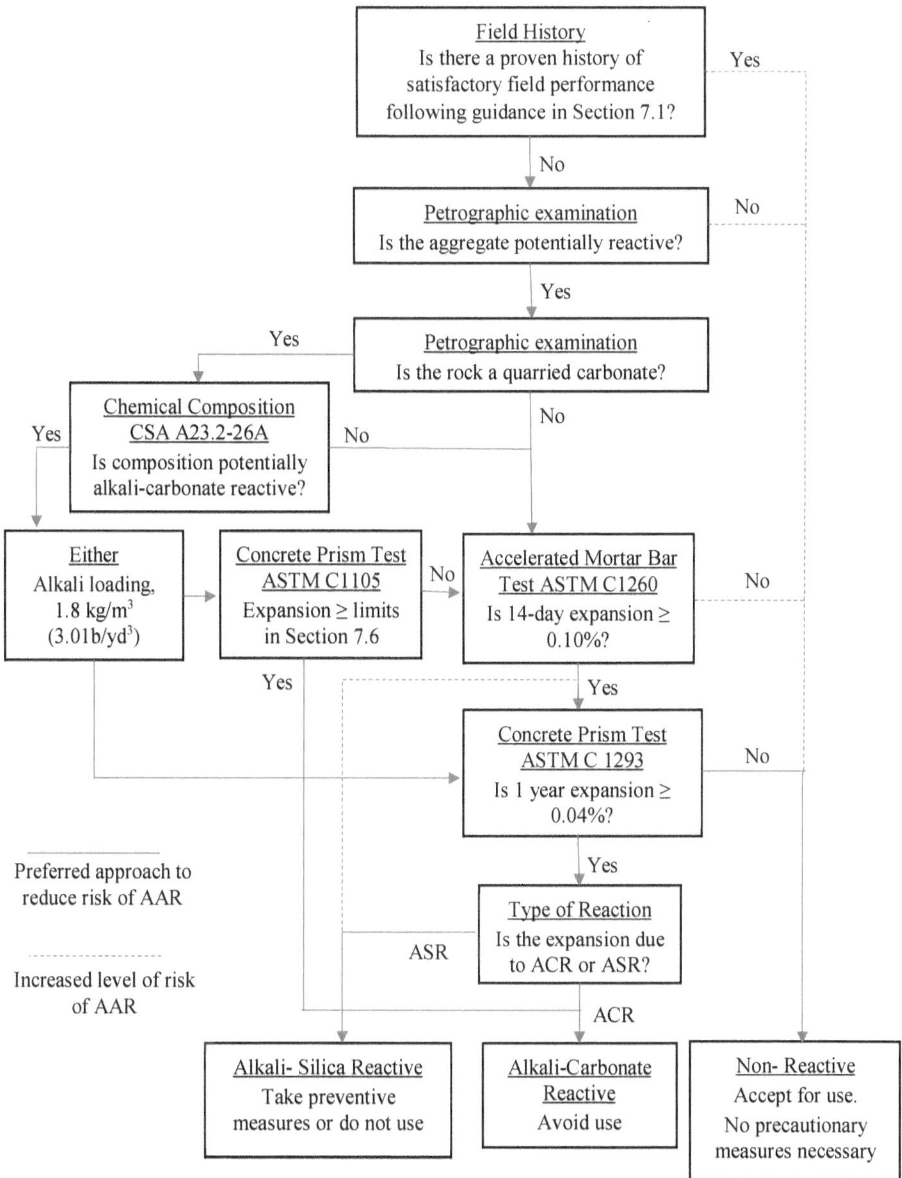

Figure 32: Sequence of laboratory tests for evaluating aggregate reactivity [(ASTM C1778 (2014)] (Nixon & Fournier, 2017)

4.2.1 Petrographic Examination

This method is used to identify the siliceous phases that exist in a specific aggregate. It depends on the assessor having prior knowledge of the types of silica that are reactive in that specific region, which is achieved by studying concrete made with aggregates from that region and that has been damaged by AAR. This method also depends on the petrographer being skilled and experienced. Petrographic examination has been used as a basis of specification in countries such as Denmark, Norway and Sweden, and is used to quantify the proportion of reactive particles in an aggregate (Nixon & Fournier, 2017). Despite this, studies like the one carried out by the PARTNER Project showed that interlaboratory agreement was very poor, especially in cases where the assessors were not familiar with certain types of aggregates.

RILEM produced the most comprehensive petrographic methodology, which was given the designation AAR -1.1 (Nixon & Sims, 2016). RILEM then also produced a petrographic atlas of reactive aggregate types, which can help petrographic assessors with examining rock aggregates that they are not familiar with. This atlas was given the designation AAR-1.2 (Nixon & Sims, 2016). Other standards bodies have also produced guides for petrographic examination, such as ASTM C295-12 (2014) and BS 812-104 (1994). ASTM C295-12 is the Standard Guide for Petrographic Examination of Aggregates for Concrete in which optical microscopy and possibly additional procedures such as X-Ray diffraction (XRD) analysis, differential thermal analysis (DTA), infrared spectroscopy, or other scanning electron microscopy (SEM) energy-dispersive x-ray analysis (EDX) (ASTM, 2003) methods are employed to identify the constituents of aggregate samples. The BS 812-104: 1994 is the Method for qualitative and quantitative petrographic examination of aggregates (BS, 1994), which is employed to examine fine and coarse aggregate samples in order to determine their petrographic composition. Since the petrographic examination method has poor precision, it is primarily used as a preliminary method to aid in choosing and determining the effectiveness of other laboratory methods (Nixon & Fournier, 2017). The petrographic examination method is therefore part of the approach recommended by RILEM in the overall assessment guide on using RILEM methods, which has the designation AAR-0 (Nixon & Sims, 2016).

The North American approach shown in Figure 32 shows that the petrographic examination is not only performed to identify potentially reactive aggregate types, but also to identify aggregates from carbonate rocks which are produced in quarrying activities. These particular aggregates go through a screening chemical method where potential alkali-carbonate reactivity is identified (Nixon & Fournier, 2017).

4.2.2 Screening With Accelerated Mortar Bar Test (AMBT) and The Role of South Africa in the Development of the Test Method

After the petrographic examination, screening using the AMBT is usually the second step in the assessment. This method is used to evaluate the potential of alkali-silica reactivity of aggregates in concrete (Nixon & Fournier, 2017). The main advantage of employing this method is that results can be obtained within a few weeks or even a few days in the extremely accelerated methods. It should however be noted that higher level of acceleration will tend to produce unreliable results, in which unreactive aggregates are typically identified as reactive (Nixon & Fournier, 2017), even though they are actually suitable to use in construction.

Recently, many standards bodies have arrived at the agreement that the AMBT is the preferred screening method. The AMBT was first developed in South Africa in the 1980s and later became standardised with various standards bodies. Such standards bodies which standardised this method are: ASTM C1260 (2014), AASHTO T303 (2008), CSA A23.2.25A (2014), AS 1141 60.1 (2014) and RILEM AAR-2 (Nixon & Sims, 2016) and the SANS 6245 (2006) Potential reactivity of aggregates with alkalis (accelerated mortar prism method).

The rapid mortar bar test method was the developed and first published by Dr. Bertie Oberholster and Davies in the 1980s at the NBRI (National Building Research Institute) of the Council for Scientific and Industrial Research (CSIR) in South Africa (Hooton & Rogers, 2003). Oberholster and Davies made use of the AMBT method to assess the effectiveness of varying the proportions of different supplementary cementitious materials to prevent concrete expansion due to AAR. In the mid-1990s, after Oberholster and Davies developed this method, the Australian AMBT methods were developed and published, and the ASTM then also developed and adopted an AMBT method as well, which was given the designation ASTM C1260. Other standards bodies also standardized this method. These included AASHTO T303 (2008), CSA A23.2.25A (2014), AS 1141 60.1 (2014) and RILEM AAR-2 (Nixon & Sims, 2016) and the SANS 6245 (2006) Potential reactivity of aggregates with alkalis (accelerated mortar prism method). All these methods were based on the original work of Oberholster and Davies (Shayan & Freitag, 2017). South Africa played a fundamental role in the development of the accelerated mortar bar test, as all similar tests which followed were based on the original test done at the NBRI.

This method entails immersing mortar bars cast with the graded aggregate in a 1N sodium hydroxide solution at 80°C for approximately 14 days, depending on the version implemented. Any expansions of the mortar bar are then monitored. This test has been proven to give positive inter-laboratory trials by RILEM as well as the EU "PARTNER" project (Nixon & Fournier, 2017) and there has been a good correlation between trial results and the experience in practice. For non-reactive aggregates, the expansion limits are generally between 0.1% and 0.15% depending on the version of the test that is implemented. It has been found however, that the method is not reliable when testing porous flint aggregates and practical examples from Argentina have shown that the test does not detect some slowly reactive aggregates when the standard criteria are applied. For that reason, the expansion limits in the method are applied at 10 and 21 days to identify slowly reactive aggregates in Australia (Nixon & Fournier, 2017).

In North America, it was observed that the results from the AMBT are in partial agreement in some cases with results yielded using ASTM C1293-8b (2014) tests or from practical experience. There have also been many aggregates that have given wrong or misleading results when using this test, particularly when compared to the concrete prism test which yields more realistic and accurate results (Nixon & Fournier, 2017). Figure 32 shows the North American testing scheme, which recommends that quarried carbonate aggregates should undergo a quick chemical method, in accordance with CSA A23.2-26A (2014) before they are tested according to AMBT. Doing this test has proven that some argillaceous dolomitic limestone aggregates in Canada that have a history of being associated with alkali-carbonate reactivity have exhibited extensive cracking in structures but they pass the AMBT (Nixon & Fournier, 2017).

4.2.3   Concrete Prism Test (CPT)

The Concrete Prism Tests are longer term tests which are generally regarded as the laboratory tests and which are most likely to give an accurate reflection of the behaviour in real concrete structures (Nixon & Fournier, 2017). Many standards bodies such as the RILEM (with the test designated AAR-3) (Nixon & Sims, 2016), ASTM C1293-8b (2014), CSA A23.2-14A (2014), BS 812-123 (1999), Norwegian Concrete Association (2005) (in Norway), Deutscher-Ausschuss fur Stahlbeton (2001) (in Germany) and AS 1141 60.2 (2014) have standardised CPTs. In all these standards, the most common specimen size is 75x75x250-300 mm, but in the German and Norwegian tests, bigger test specimens are used, for example in the Norwegian CPT method, specimen size used is 100x100x500 mm. Also, generally in CPTs, a high content (420 - 440 kg/m^3) of high alkali cement (about 0.8-1.3% Na_2O_{eq} depending on the version of the method implemented) is used to reach a high level of alkali in concrete (Nixon & Fournier, 2017). In some methods such as the ASTM C1293-8b (2014), CSA A23.2-14A (2014), AS 114 60.1 (2014), the level of alkali is increased to 1.25% Na_2O_{eq} by mass of cement by dissolving NaOH pellets in the mix water (Nixon & Fournier, 2017). Thereafter, the specimens are stored at an elevated temperature of typically 38°C and a high humidity. The specimens are then monitored for one year and any expansion is noted. The pessimum effects are monitored by varying the quantity of test aggregate. The expansion limits are between 0.03% and 0.05% for non-reactive aggregates, but they vary depending of local experience (Nixon & Fournier, 2017). It should be noted that these expansion limits are dependent on which CPT standard is being followed.

A negative aspect of these CPTs was found to be the difficulty in balancing between ensuring a sufficient amount of humidity and minimising the loss of alkali from the specimen through leaching (Nixon & Fournier, 2017). In the earlier application of these methods, it was found that wrapping specimens in damp cloth of placing them in fog rooms produced excessive leaching and called into question the accuracy of the results. The consensus is now for specimens to be stored unwrapped in airtight containers above water. RILEM and the EU "PARTNER" conducted inter-laboratory trials which found that there was good agreement between the tests and field practice for most aggregates, even though identifying slowly reacting aggregates may require longer test periods or lower acceptance criteria (Nixon & Fournier, 2017). PARTNER found reasonable precision, although Fournier *et al.* (2012) showed that there is a high multiple-laboratory variability, due to different factors such as the use of non-reactive fine aggregates together with the coarse aggregate used in the test (Nixon & Fournier, 2017). In response to this, RILEM developed an accelerated CPT, which was given the designation AAR-4.1 (Nixon & Sims, 2016). The AAR-4.1 was based on the French Performance Test (AFNOR NF P18-454, 2004) (Nixon & Sims, 2016). In this test, the concrete specimens are stored at a high temperature of 60°C in a high humidity reactor for a test period of 15 weeks (Nixon & Fournier, 2017). A good correlation was found between this test and those performed at 38°C. Due to the reasonable precision found in the PARTNER trials, it was established that this test method was more effective in the identification of slowly reacting aggregates (Nixon & Fournier, 2017).

It should be noted that the main difference between this CPT and the performance tests discussed in a later section is that the CPTs specify the concrete mix ratios, depending on which standard is applied. However, in the concrete performance tests, the test specimens are made with the concrete mix design that is planned to be used in the actual construction and therefore the results represent the concrete performance of the concrete proposed for the construction.

4.3 Concrete Performance Test with Specific Concrete Mix

Concrete performance tests are reliable performance tests of the actual concrete mix that is going to be used in the construction of the concrete structure and then making a decision about whether the material used in the mix are suitable for use (Nixon & Fournier, 2017). The difficulties with these tests lie in their ability to achieve adequate acceleration to provide answers in a useful timescale while at the same time giving a reliable assessment of their performance. These difficulties have thus limited the development of performance tests. AFNOR NF P18-454 (2004) standardised a method which is based on monitoring the expansion of the concrete test prisms which are stored at 60°C, and which is the basis of RILEM AAR-4.1 discussed previously (P. Nixon & Fournier, 2017).

In France, this test is used to establish the threshold level of alkalis required for a damaging reaction to occur in concrete, and then provide an opportunity to reduce the level of alkali in the actual mix to be used by a safety factor (Nixon & Fournier, 2017). For years now, RILEM has been developing a similar performance test and currently, the consideration is for the test to be based on AAR-3 (P. Nixon & Fournier, 2017). There is currently a proposed RILEM technical committee (TC) document with the designation TC 258-AAA which has still not been finalised.

4.4 AFNOR P 18-588 Microbar Test- Switzerland

In Chapter 2, eight tunnels in Switzerland were discussed. These tunnels showed signs of AAR in the form of typical crack patterns, pop outs and excessive deposits on the concrete surface. These tunnels were assessed to verify the existence of AAR damage and to determine the extent of the damage (Leemann et al., 2005).
To assess AAR damage, visual assessments were conducted first. These assessments showed signs of typical crack patterns, pop outs and excessive deposits on the surface, which indicated AAR.

The AFNOR P 18-588 microbar test was conducted to measure the potential alkali aggregate reactivity. In the case of the Swiss tunnels, this method showed a good correlation, with $R^2 =$ 0.94 (for 20 samples) with the NBRI (National Building Research Institute) test (Leemann et al.,

2005). This test should therefore give results which are comparable to AMBT tests such as the ASTM C1260 as well as the RILEM method AAR-2 (Leemann et al., 2005). The AFNOR test uses cement to aggregate (c/a) ratios of 2, 5 and 10. In Switzerland, there are no known aggregates with pessimum behaviour and therefore the value obtained from using the c/A ratio of 2 is used in the classification of the aggregates. The classification of the reactivity of the aggregates is divided into three groups: not reactive (NR), reactive (R) and strongly reactive (SR). These classification groups are in accordance with the sample expansion as follows:
- NR: expansion $\leq 0.1\%$
- R: expansion $> 0.1\% < 0.2\%$
- SR: expansion $\geq 0.2\%$

Further details of this research are shown in Appendix 6.

4.5   Concrete Imaging – Canada

A research was carried out to analyse surface AAR map-cracking in bridge decks in Eastern Canada (discussed in Chapter 2 of this book), using thermal, colour and greyscale imagery. The concrete deterioration features that were characterised and measured were: total amount of cracking, total length of cracks, as well as the range of crack widths. These concrete deterioration features were characterised and measured with the use of texture analysis and an artificial neural network. These concrete imaging methods are based on non-destructive testing (NDT) techniques and their use makes it possible to efficiently obtain comprehensive information regularly without disrupting traffic. This research included concrete specimens which were obtained from the field, as well as concrete blocks and slabs prepared in the laboratory, all of which experienced different levels of alkali aggregate reaction induced surface cracking (Kabir, 2006).

Field Specimens

Components from various bridges with varying levels of AAR damage were used in this study. Figure 33 shows the AAR damaged bridges in the St Lambert Lock Bridge in Montreal and a train bridge located in Sherbrooke, where components were obtained from. AAR results in swelling and eventual cracking in concrete, and the level of swelling gives an indication of the concrete deterioration, loss of rigidity and decreased mechanical properties. The components obtained from the St. Lambert Lock Bridge were severely affected by AAR, and exhibited various rates of concrete swelling. Components retrieved from bridges in Sherbrooke, including the train-bridge, all exhibited different levels of concrete deterioration.

Figure 33: (a) and (b) show AAR induced map cracking in the bridge component at the St. Lambert Lock in Montreal. (c) is the train-bridge beam in Sherbrooke (Kabir, 2006)

Laboratory Specimens

Two sets of specimens with various concrete mix designs were produced in laboratories to be used in the research. The first set was made up of three concrete blocks, of size 400 x 400 x 700 mm, which were cast with a coarse aggregate of reactive limestone. These blocks were left exposed to weather elements at the CANMET site in Ottawa, Canada. The second set comprised three concrete slabs of size 1000 x 1000 x 250 mm (Kabir, 2006). These slabs were given the designation D1, D2 and D3. Slab D1 was cast with a non-reactive aggregate and sets D2 and D3 were cast with reactive limestone (Kabir, 2006). After batching and wrapping the slabs in damp terry cloths, they were then stored in ambient air at a temperature of 20 ±2°C at the GRAI laboratory at the University of Sherbrook in Québec, Canada.

Tests of the amount of cracking were carried out on specimens at both laboratories. This was done due to the fact that the amount of cracking is closely linked to the level of expansion of the concrete as well as concrete deterioration indicators such as a decrease in concrete mechanical properties (Kabir, 2006). The velocities of compression (P) waves decrease when the amount of concrete damage increases. In these tests, the P-wave velocities were measured by employing the impact-echo method. The expansion of the specimens was measured by fixing stainless steel studs on the top surfaces as well as the sides of the specimens. Table 15 shows the concrete mixes, the P-wave velocities as well as the average expansion levels measured for the test specimens (Kabir, 2006).

Table 15: Concrete mixtures proportions, average P-wave velocities and expansion measurements of CANMET and GRAI specimens (Kabir, 2006)

CONCRETE MIXTURES	CANMET			GRAI		
	D1	D2	D3	D1	D2	D3
Density (kg/m^3)	2303	2303	2317	2223	2326	2340
Cement Content (kg/m^3)	423	423	425	210	390	390
Total Na$_2$O$_{eq}$(kg/m^3)	1.69	3.81	5.31	3.81	3.25	5.25
w/c	0.42	0.42	0.42	0.75	0.66	0.66
Test Measurements						
Average P-wave velocities (ms^{-1})	4909	4513	4402	3810	3590	3440
Average expansion (%)	0.025	0.374	0.383	0.000	0.060	0.100

From the results, the following can be seen:
- Greyscale imagery performed fairly well for the CANMET blocks, with an overall classification accuracy range of 72.3% - 76.5%
- Greyscale imagery performed fairly well for the GRAI blocks, with an overall classification accuracy range of 68.7% - 75.3%
- Visual colour imagery performed slightly worse than Greyscale imagery for the CANMET blocks, with an overall classification accuracy range of 71.4% - 75.2%
- Visual colour imagery performed slightly better than Greyscale imagery for the GRAI blocks, with an overall classification accuracy range of 70.9% - 72.0%
- Thermographic imagery produced the highest overall classification accuracies for the CANMET blocks, with an overall classification accuracy range of 74.5% - 76.3%
- Thermographic imagery produced the highest overall classification accuracies for the GRAI blocks, with an overall classification accuracy range of 75.6% - 76.9%

The classification results from all the laboratory specimen show that infrared thermography performed better than both the visual colour and greyscale imagery, and therefore the results of infrared thermography were used to quantify the different levels of AAR damage of the different specimens. Table 16 shows the results from the infrared thermography for the CANMET and the GRAI samples. The narrow and wide cracks from Figure 33, together with the expansion levels from Table 15 , for the CANMET and GRAI samples are shown in Figure 34 . The average expansion levels were multiplied by of 100 for ease of comparison.

For further analysis of the concrete surface damage, the classified thermographic images were converted into binary images, whereby the image is simplified by assigning pixels which represent the damage in the concrete a value of 1 (black) and by assigning the background pixels a value of 0 (white) (Kabir, 2006). Manual or automated methods were then employed to add the pixels in order to calculate the total length of the wide crack, and also the average width of the wide crack. The total length of wide cracks was measured by adding all the pixels along the length a specific branch of the cracks, and this total was multiplied by the pixels resolution, which was 0.26 mm. The results in Figure 34 were obtained for wide cracks (Kabir, 2006):

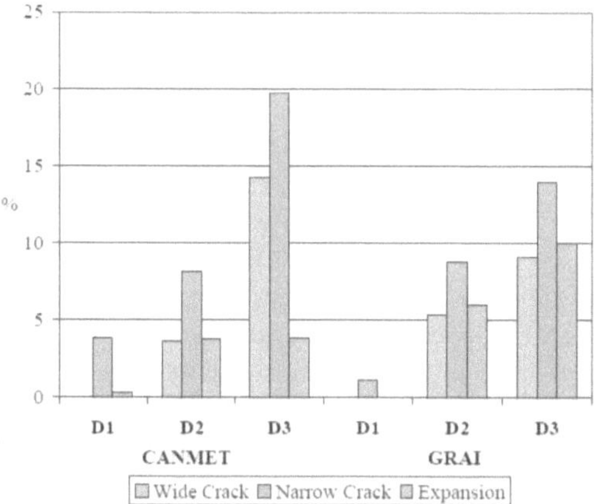

Figure 34: Comparison of concrete crack damage and expansion levels (shown in percentage on the vertical axis) among the CANMET specimens and the GRAI specimens (Kabir, 2006)

Table 16: Crack lengths and average crack widths (Kabir, 2006)

CRACK PROPERTIES	CANMET			GRAI		
	D1	**D2**	**D3**	**D1**	**D2**	**D3**
Total Crack Length (mm)	0	97.6	237.4	0	38.6	107.3
Average Crack Width (mm)	0	0.8	1.6	0	0.3	0.8

These results were consistent with the data shown in Table 15, for the CANMET and GRAI specimens. From Figure 34, CANMET specimen D1 exhibited the smallest expansion due to the fact that the concrete mix had a low alkali content. CANMET specimen D3 exhibited the biggest expansion due to the fact that the concrete mix had the highest alkali content. CANMET specimen D3 also exhibited the highest values of total length of wide cracks as well as average crack length, which corresponded with it having the lowest P-wave velocities, indicating that it had the highest level of deterioration (Kabir, 2006).

GRAI specimen D1 exhibited no wide cracks, and this corresponded with the fact that it had the lowest expansion level and thus very little damage. GRAI specimen D2 exhibited a higher expansion level, while GRAI specimen D3 exhibited the highest expansion level from the GRAI specimens. Figure 34 also shows that there is a strong correlation between the amount of wide crack damage and in the GRAI concrete specimens and the average level of expansion (Kabir, 2006).

Field Samples
Classification results were obtained as follows (Kabir, 2006):
- Thermographic, colour, and greyscale images were taken from bridge components of St. Lambert Lock

- Colour and greyscale imagery were taken of Jacques-Cartier, the Joffre and Terrill bridges and the train bridge.

The results obtained were similar to those obtained from the CANMET and GRAI classification. As with the laboratory specimens, infrared thermography also exhibited higher accuracies than greyscale and visual colour imagery for the bridge component of the St. Lambert Lock. For the other field samples, colour imagery performed better than greyscale imagery. As with the laboratory specimens, the wide crack lengths and the crack widths were calculated (Kabir, 2006). From all the field samples, the Joffre Bridge component exhibited a wide range of crack widths, ranging from 0.15 mm to 0.3 mm (Kabir, 2006). The classified colour image of the component from Joffre Bridge is shown in Figure 35. Zoomed areas indicate the different ranges of cracks widths. The crack width ranges are as follows:
- C1: 0.1mm – 0.15mm
- C2: 0.15mm – 0.20mm
- C3: 0.20mm – 0.30mm
- C4: above 0.3mm

It can be concluded that making use of the GLCM (grey level co-occurrence matrix) texture method and the ANN (artificial neural network) classification technique is effective in analysing surface damage due to AAR deterioration in concrete. The application of these approaches makes it possible to not only detect surface deterioration in concrete such as cracks, but also the quantification of detected defects using thermographic, visual colour and greyscale (Kabir, 2006). The classification accuracies produced by thermographic imagery were the highest for all the samples where this imagery was employed, and it therefore produced the best results. Colour imagery also produced acceptable results and it was employed in analysing damage in some field samples (Kabir, 2006).

Another conclusion that may be drawn was that there was a good correlation between the amount of wide cracks in the concrete sample and its expansion level. These methods used are less costly and less time-consuming than the conventional visual inspection methods, and therefore allow for evaluations to be carried out more frequently to supplement the conventional visual assessments (Kabir, 2006). Another advantage of employing these methods is that they allow for quantitative evaluations, such as assessing the total amount of surface damage in the images. This can then improve the quality of the concrete condition information which is obtained from conventional inspections that are performed with the aim of making decisions regarding maintenance and repairs of concrete structures (Kabir, 2006).

Further details of this research are shown in Appendix 7.

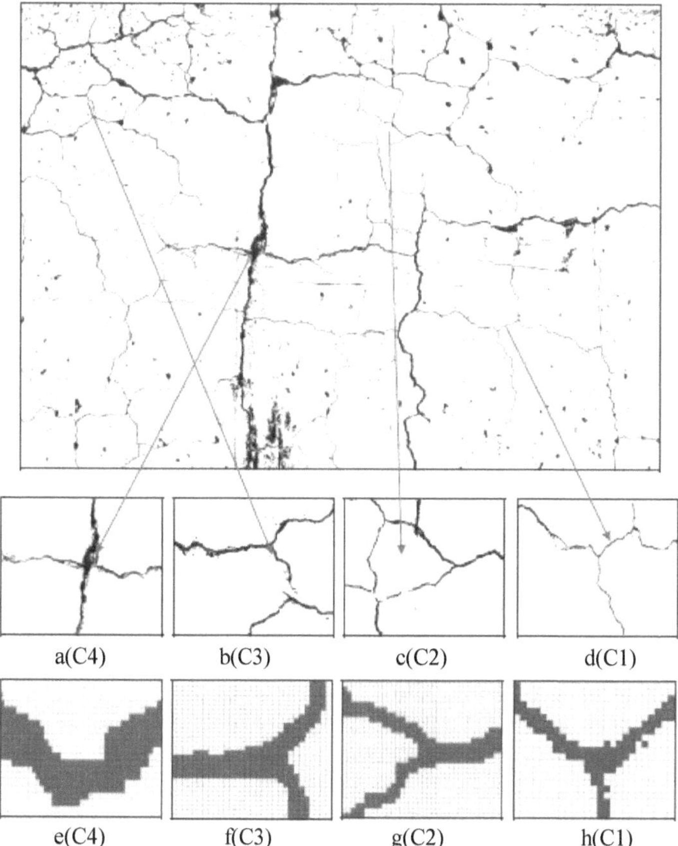

Figure 35: Different ranges of crack widths from binary image of Joffre Bridge. (a)-(d) zoomed to 1x and (e)-(h) zoomed of 20 times with a grid of 1 pixel per square (Kabir, 2006)

4.6   Stiffness Damage Test (SDT) and Damage Rating Index (DRI)

Laboratory tools have been developed to test either the actual degree of ASR damage in concrete or the potential further damage that can be caused by ASR. Two of those tools are the Stiffness Damage Test (SDT) and the Damage Rating Index (DRI), and these have been seen to present great potential for application in practice (Sanches et al., 2016). ASR-affected concrete was evaluated making use of SDT and DRI. The evaluations were performed on three different grades of concrete – 25, 35 and 45 MPa, using two types of highly reactive aggregates, namely the Texas sand and New Mexico gravel. This section outlines an evaluation of the applicability of SDT and DRI in detecting and quantifying the damage caused by ASR as a result of using deleterious coarse or fine aggregate in the concrete mix (Sanchez et al., 2016).

4.6.1   The Stiffness Damage Test (SDT)

The Stiffness Damage Test (SDT) has been used since the 1980s by Crisp and Co-workers (Crisp et al., 1993), to quantify damage caused by ASR. It was reported that there was a good relationship between the crack density and loading/unloading cycles (i.e. the stress/strain relationship) of rock specimens. From this, the SDT was then proposed, and it was based on the cyclic loading in compression of concrete cylinders or cores with a diameter greater than 70 mm and a length/diameter of 2 - 2.75. At first, the SDT test involved applying stress up to 5.5 MPa at a rate of 0.1 MPa/s. However, the testing procedure needed to be made non-destructive so that the specimens could be used for further tests. This led to the loading being controlled by a microprocessor and then being repeated five times.

More than one thousand tests were performed on cores which were extracted from concrete structures which were damaged. The stress-strain response was analysed and several parameters were identified to diagnose the extent of the damage in a specific specimen. These were the modulus of elasticity; hysteresis area and non-linearity index (Sanchez et al., 2016). After carrying out several evaluations on the reliability of SDT using various reactive rocks at various expansion levels, it was found that the parameter that gave the best output response for SDT was the hysteresis area (Sanchez et al., 2016). The hysteresis area H, measured in J/m^3 is the area of the hysteresis loops, averaged over the last four cycles, due to the fact that concrete with damage showed greater energy loss (or hysteresis areas) compared to concrete that has not been damaged (Sanchez et al., 2016). The hysteresis area particularly gave good outputs for the initial cycle of the test specimens which were loaded to a maximum of 10 MPa, as lower stress levels did not allow the ASR induced microcracks to stress enough to make it possible to draw reliable information regarding the level of expansion due to ASR in the concrete that was tested (Sanchez et al., 2016).

Eventually, Sanches et al. did further studies on SDT by casting 25, 35 and 40 MPa concrete using different coarse versus fine reactive aggregates. This was done with the aim of verifying the impact of the test loading or other input factors such as the concrete environment, humidity, specimen size or the test loading on the output test analyses (Sanchez et al., 2016). The SDI was conducted by subjecting concrete test cylinders to five cycles of loading/unloading at a loading rate of 0.1 MPa/s. The specimens were then tested at a strength level which was equal to 40% of the 28-day strength of the concrete design strength.

Two highly reactive aggregates (New Mexico gravel (NM) and Texas sand (Tx)) were used in the study. Non-reactive coarse and fine aggregates were used in combination with the reactive aggregates to make the concrete (Sanchez et al., 2016). The concrete mix designs are shown in Table 17. The number shown in brackets represents the volume that the materials occupy, given in L/m³.

Table 17: Concrete mix designs used in the tests (Sanchez et al., 2016)

Ingredients	25 MPa- Materials (kg/m³)[1]		35 MPa- Materials (kg/m³)		45 MPa- Materials (kg/m³)	
	Texas sand	NM gravel	Texas sand	NM gravel	Texas sand	NM gravel
Cement	314 (101)	314 (101)	370 (118)	370 (118)	424 (136)	424 (136)
Sand	790 (304)	714 (264)	790 (304)	714 (264)	790 (304)	714 (264)
Coarse aggregate	1029 (384)	1073 (424)	1029 (384)	1073 (424)	1029 (384)	1073 (424)
Water	192 (192)	192 (192)	174 (174)	174 (174)	157 (157)	157 (157)

Two indices were used in these tests, namely the SDI and the PDI (Plastic Deformation Index). The SDI represented the dissipated energy over the five compression cycles, or the total energy that was implemented in the system (obtained by calculating the area under the stress/strain graph). The PDI represented the plastic deformation, also over the five compression cycles (Sanchez et al., 2016). The results of these two indices are shown in Figure 36 and Figure 37.

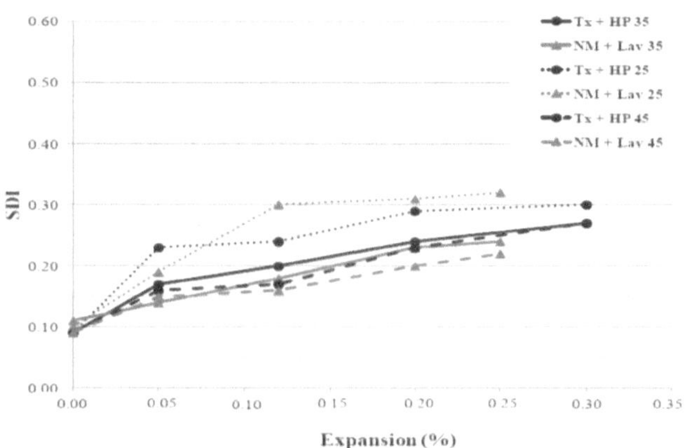

Figure 36: Stiffness damage indices for 25, 35 and 45 MPa concrete mixes - SDI results (Sanchez et al., 2016)

From the SDI results in Figure 36, it can be seen that the SDI able to distinguish very well between the different levels of expansion (or the different amounts of internal cracking) for the different concrete mixes, aggregates and strengths. Depending on the concrete strength, there is generally a concave or linear relationship of the SDI against the expansion level of the concrete specimens.

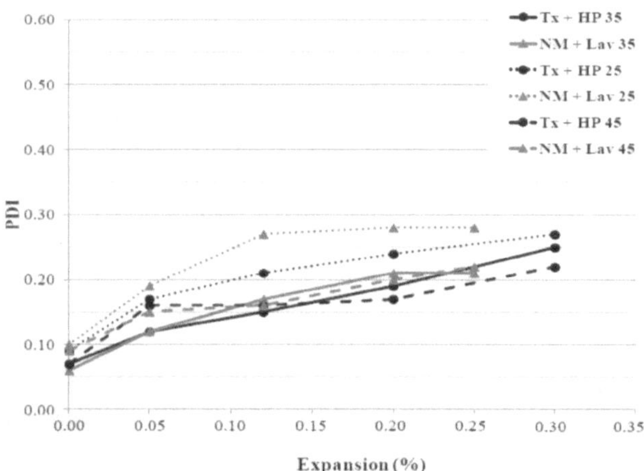

Figure 37: Stiffness damage indices for 25, 35 and 45 MPa concrete mixes - PDI results (Sanchez et al., 2016)

From the PDI results in Figure 37, it can be seen that, just like the SDI, the PDI is also able to distinguish very well between the different levels of expansion (or the different amounts of internal cracking) for the different concrete mixes, aggregates and strengths. There is also generally a concave or linear relationship of the SDI against the expansion level of the concrete specimens.

The output test responses were also evaluated against the expansion of the test specimens. The findings from the SDI evaluations were as follows (Sanchez et al., 2016):

i. It is best to carry out SDT using a percentage of concrete strength and not a fixed load,

ii. Using up to 40% of the concrete strength distinguishes different damaged concrete specimens,

iii. If 40% of the concrete strength is used, this makes it possible to use the same concrete sample to perform other analyses like the compressive and tensile strengths. Specimens maintain their non-destructive nature up until that point.

iv. Hysteresis area was the one parameter that was found to give the best output results.

v. The SDT output analysis is also influenced by the concrete's curing history, the size and geometry, the concrete sample's environment and also any tests which were performed on the samples in the past.

vi. Using indices such as SDI and PDI considers the ratio of dissipated energy to total energy in the system and thus gives a better representation of the actual damage of the affected concrete materials.

4.6.2 Damage Rating Index (DRI)

Damage Rating Index (DRI) is a microscopic analysis which is carried out using a stereomicroscope of about 15x magnification, and the features caused by ASR damage are counted through a 1 cm^2 grid that is drawn on the surface of a polished concrete section (Sanchez et al., 2016). The number of counts that corresponds to the different petrographic features is multiplied by weighting factors, which balance their relative importance to the damage mechanism (in this case ASR) (Sanchez et al., 2016). Even though the multiplication factors were logically determined, they are relatively arbitrary. For the DRI analysis, the surface should ideally be 200 cm^2, or even more in cases where mass concrete with larger aggregates is used (Sanchez et al., 2016). For comparison though, it has been normalised that the largest DRI value is 100 cm^2 and the results are usually presented with the use of charts to better visualise the various damage features of the specimen being studied.

The study conducted by Sanches et al. involved the following materials (Sanchez et al., 2016):
- Three concrete mixes of design strengths 25 MPa, 35 MPa and 45 MPa.
- Two highly reactive aggregates, namely Texas sand and New Mexico gravel. The coarse aggregate had a size range of 5-20mm.
- The two highly reactive aggregates were used in combination with non-reactive coarse and fine aggregates for concrete mixing.

  The DRI test was conducted by casing all the concrete mixes for the expansion levels predetermined for the study. Counts of cracking in the aggregate particles were made until the size of 1 mm.

  Figure 38 shows the DRI results in terms of the relative importance of the different petrographic deterioration features.

The following observations can be made from the DRI results (Sanchez et al., 2016):
- In all the polished sections, a common feature of deterioration corresponds to closed cracks within the aggregate particles (CCA). This feature is even more apparent in the Tx mixes;
- A progressive increase can be seen in the amount of cracking in both aggregate particles (OCCA and CCAG) and the cement paste with and without gel (CCP and CCPG), with an increase in expansion of the concrete specimens;
- For all the concrete mixes and all the aggregates, the DRI value corresponds very well with the levels of expansion measured. Additionally, the DRI values don't seem to be affected by the strength of the concrete;
- The DRI values don't seem to be affected by whether the deleterious expansion is caused by fine aggregate (Tx) or the coarse aggregate (NM);
- For the 45 MPa concrete mixes, it could be seen that higher DRI values (compared to 25 and 35 MPa) were obtained at lower expansion levels, and they remained stable up to about 0.12%, after which they increase almost linearly like the other mixes;
- The data shows that there is existing damage in the test specimens (in the DRI range between 100 and 140 from cracking in the aggregate and cement paste) for concrete

specimens that show no significant level of expansion (shown as 0% in Charts (b) and (e)).

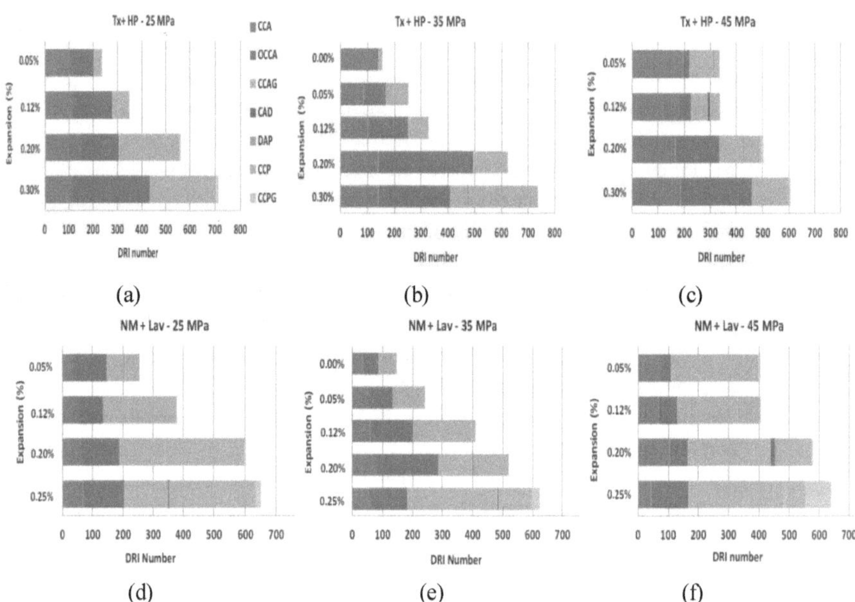

(a)　　　　　(b)　　　　　(c)

(d)　　　　　(e)　　　　　(f)

Figure 38: DRI results for the concrete mixes in the study. Texas sand in charts (a), (b) and (c). NM gravel + Lav in charts (d), (e) and (f). Legend is shown in Chart (a) (Sanchez et al., 2016)

The results from these SDT and DRI tests showed the following (Sanchez et al., 2016):

1. Both the SDT and DRI are able to quite accurately distinguish different levels of ASR expansion in the concrete. These expansion levels illustrate important mechanical and microscopic information about either the extent of cracking or the cracking features from the ASR damage.
2. The mechanical and microscopic data correlates very well with the expansion levels that were studied.

4.7 Multi-Physics Approach (Non-Linear Acoustic Measurements and Microwave Materials Characterisation Measurements)

One of the most recognisable feature of Alkali Silica Reaction is that it exudes a gel to the surface of the concrete. Once the gel has formed and made its way to the surface, extensive expansion and damage have taken place, creating unique challenges for repairing such structures. Additionally, at that point, deleterious aggregates may already have been used in subsequent constructions. It would therefore be greatly beneficial to detect ASR within concrete early in order to significantly contribute to the sustainability of concrete structures. The multi- physics approach helps to study ASR comprehensively by linking chemical changes (such as the transition of water from the pore solution to the gel), physical changes (such as micro-cracking and the formation of gel) and mechanical properties (such as changes in material linear elasticity or increasing non-linearity with increasing damage) (Rashidi et al., 2016). The multi-physics approach for assessing ASR provides an opportunity to understand the nature of the ASR, and can thus be used to develop techniques and tools to assess and monitor ASR in existing concrete structures (Rashidi et al., 2016). This approach is a combination of two non-destructive evaluation techniques, which are (Rashidi et al., 2016):

- Non-linear acoustic measurements, which are sensitive to micro-cracking; and
- Microwave materials characterisation measurements, which are sensitive to moisture. This includes the transition of water from its free state in the concrete pore solution to the bound state that it is in within the ASR gel.

When compared with the expansion assessment and the damage index rating obtained from petrographic analysis on standard mortar bars, a correlation can be seen between all the measures. Rashidi et al. conducted an experiment to test the applicability of the multi-physics method. The experiment involved the preparation of standard mortar bars using three aggregate sources from the United States of America, two of which were potentially reactive and one non-reactive aggregate. These mortar bars were then cured and placed under exposure conditions in accordance with the accelerated mortar bar test as set out in ASTM C1260 (Rashidi et al., 2016). Expansion, petrographic, microwave and non-linear acoustic assessments were then carried out on the samples.

In order to do expansion measurements, mortar bars were cast, and then demoulded after 24 hours. They were then stored in deionised water at a temperature of 80°C for 24 hours, after which they were submerged in 1N NaOH solution at a temperature 80°C for 14 days, during which the expansion and non-linearity were measured periodically (Rashidi et al., 2016).
In order to perform microwave measurements, two sets of samples were prepared. One set of samples was exposed to the same conditions as the mortar bars, while the other set was stored in deionised water at a temperature of 80°C after demoulding. The purpose of this was to compare the water-exposed and sodium hydroxide-exposed sets in order to differentiate the effects of cement hydration and high temperature curing in an alkaline solution from the effects of ASR (Rashidi et al., 2016).

Petrographic analysis in the ASR-affected concrete samples was carried out using the damage rating index (DRI). The DRI assigns weights to different types of defects in the concrete, as shown in Table 18 (Rashidi et al., 2016).

Table 18: Defect scaling factors used for DRI (Rashidi et al., 2016)

Defect Type	Scaling Factor
Crack in coarse aggregate filled with gel	2
Crack in coarse aggregate without gel	0.25
Crack in cement paste filled with gel	4
Crack in cement paste without gel	2
Reaction rim	0.5
Air void with gel	0.5

The DRI was performed by cutting a 20-30 mm thick concrete section from a cylinder or prism sample, polishing it and then drawing grid sizes of 15 x 15 mm^2 on the surface of the sample. The number of defects in each grid was then counted at the magnification of 16x and then multiplied by its weighting factor. The answers were added together and the final answer was reported per 100 cm^2 (Rashidi et al., 2016)·

Microwave measurements were performed by placing a mortar sample into a waveguide and then connecting both ends of the waveguide to calibrated ports of an Agilent 8510C vector network analyser (VNS) and then recording the transmission coefficient S_{21} and reflection coefficient S_{11} of a signal, and the relative permittivity and loss factor were then calculated (Rashidi et al., 2016). Due to the fact that S_{11} and S_{21} both interact with the sample, the calculated values were sensitive to the physical and chemical properties of the sample, such as the presence and state of water in the mortar (Rashidi et al., 2016). The test setup is shown in Figure 39.

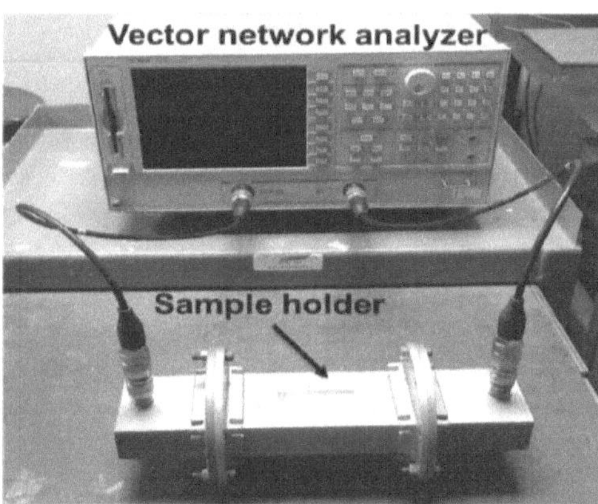

Figure 39: Test setup for the microwave measurements (Rashidi et al., 2016)

Nonlinear impact resonance acoustic spectroscopy (NIRAS) measurements were performed by applying small and incremental forces to the midpoint of the sample and then capturing the behaviour of the sample using an accelerometer which was located at the end of the sample (Rashidi et al., 2016). The transient vibration response of the sample was measured in the time domain using an oscilloscope, shown in Figure 40 (Rashidi et al., 2016). The sample responses to the impacts were then analysed in the frequency domain.

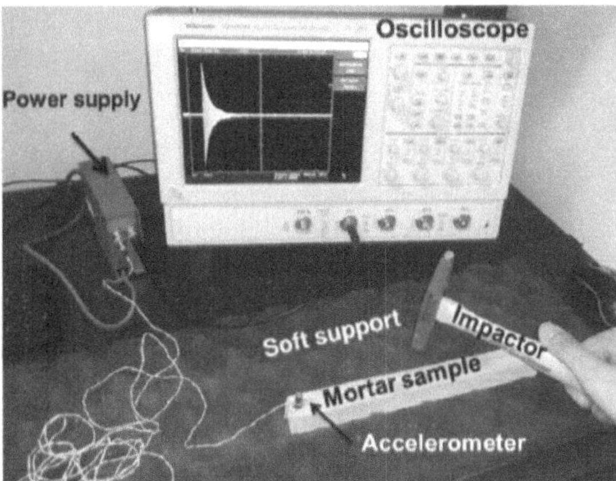

Figure 40: NIRAS test setup (Rashidi et al., 2016)

All the experiments which were performed (expansion, petrography, microwave measurements and NIRAS) were able to successfully differentiate the potentially deleterious aggregates from the innocuous aggregates in the mortar samples which were tested (Rashidi et al., 2016). Although all 4 experimental results show correlation, the strongest correlation exists between the expansion and the non-linearity and the dielectric properties and the damage evaluated using the petrographic analysis. The following conclusions were drawn from the experiments which were conducted (Rashidi et al., 2016):

- Relative permittivity shows higher sensitivity to the formation of ASR gel and the samples which showed higher relative permittivity were those cast with potentially reactive aggregate when compared to the samples cast with innocuous aggregates.
- The DRI values increased with an increase in time of the exposure to the accelerated mortar bar test (AMBT). These values showed a consistent increase with the expansion data for the samples cast with potentially reactive aggregates, while this was not the case with samples prepared with innocuous aggregates.
- There was a strong correlation between the cumulative average non-linearity parameter and the expansion of each sample type. This correlation is stronger for the samples cast with potentially reactive aggregate compared to the samples cast with innocuous aggregate.
- There was a stronger correlation between the cumulative average non-linearity parameter and the DRI for samples cast with potentially reactive aggregate than those cast with innocuous aggregate.
- There was an agreement of the relative permittivity and DRI of sample types, and this agreement may stem from the larger weights which have been assigned to the defects

which contain ASR gel as well as how sensitive of the relative permittivity is to the presence of gel.

To summarise, through these experiments, it has been established that there is a connection between gel formation, damage and expansion. The combination of non-linear acoustics and microwave measurements provides better information about the evolution of microcracks and the formation of gel during ASR expansion (Rashidi et al., 2016). This experiment was able to connect gel and damage in a direct manner using an accelerated test, however, it is possible that the composition and expansivity of the gels encountered in the field could or could not be the same as those formed during these experiments (Rashidi et al., 2016). Therefore, it is necessary for further work to be done on the application of the multi-physics approach using aggregates varying in reactivity and exposed to field conditions in order to determine the relationship between the gel properties and micro-cracking and expansion.

4.8 Critical Analysis of AAR Testing Methods Discussed

This chapter discussed the following AAR testing methods:
- Integrated Schemes According to RILEM and North American Guidelines, comprising three assessment stages, namely:
  - o Petrographic examination
  - o Accelerated mortar-bar test for screening
  - o Concrete prism tests
- Concrete Performance Tests
- AFNOR P 18-588 Microbar Test used in Switzerland
- Concrete Imaging performed in Canada
- Stiffness Damage Test (SDT) and Damage Rating Index (DRI)
- Multi-Physics Approach (Non-Linear Acoustic Measurements and Microwave Materials Characterisation Measurements)

An analysis will now be done on each testing method discussed in this chapter.

The integrated schemes according to RILEM and North American guidelines provide a systematic approach to assessing the AAR potential of aggregates and these schemes are widely used all over the world. They comprise three assessment stages, namely: the petrographic examination, accelerated mortar-bar test (AMBT) for screening and concrete specimens for long term testing (concrete prism tests). Under these schemes, both short and long term testing of specimens can be performed, allowing the user an opportunity to get more accurate results when they opt for the long term tests (which may take about one year to perform). The schemes also allow for various testing standards to be applied, making them flexible for use throughout the world. The AMBT allow for results to be obtained in times as short as 14 days, which is very important in a fast-paced world where construction projects need to be completed in short periods of time. Additionally, RILEM produced a petrographic atlas of reactive aggregate types, which can help petrographic assessors with examining rock aggregates that they are not familiar with. Tests in these schemes may also be performed in isolation, and commonly, the AMBT is performed on its own without performing the petrographic examination before it and without performing the concrete prism (long term) test after it.

The concrete performance tests (CPTs) that were discussed in this chapter are tests that may be performed using the exact concrete mix design to be used in the intended construction and therefore give a good indication of how the concrete is expected to perform in operation. These tests follow the same standards as the concrete prism tests under the RILEM and North American integrated schemes and RILEM is currently busy devising a test method particularly for this purpose.

This chapter also discussed tests which assess AAR damage which has already occurred in structures in service. The importance of these tests is that they are very useful to the processes of performing condition monitoring and assessment of concrete structures and therefore form an important part of the maintenance, repair and rehabilitation of existing concrete structures.

Concrete imaging using the GLCM texture and ANN classification methods are non-destructive testing techniques and they make it possible to obtain information about structures on a regular basis without interrupting the operations of those structures. With this method, it is possible to detect the surface deterioration of structure and also to quantify the detected defects. One benefit of concrete imaging methods is that they are less costly and less time consuming than the conventional visual assessments that are performed and because of this, they allow for frequent assessments. This, coupled with the quantitative information obtained, can assist greatly with condition monitoring and inspections of concrete structures as well as the maintenance, rehabilitation and repair plans of these structures.

The Stiffness Damage Test (SDT) and the Damage Rating Index (DRI) tools are able to give the mechanical and microscopic information about the extent of cracking or the cracking features resulting from ASR damage, and these in turn give information about the expansion levels. One concern is that the SDT and DRI tests discussed in this chapter were performed on over 1000 concrete cores extracted from the concrete structures and there is no clear indication of whether the tests can be performed in a non-destructive manner.

The multi- physics approach can be used to assess ASR in existing structures by drawing a connection between gel formation, concrete damage and concrete expansion. This method detects the gel that is formed within the concrete structure, before it exudes to the surface of the concrete structure and before the visual manifestations of ASR occur. The benefit of this is that once the gel is detected within the structure and ASR is expected in that particular structure, plans can be made so that the deleterious aggregates used in the particular structure are not used in other structures, or if they must be used, appropriate avoidance measures can be employed. This method would be useful as a warning sign against using the aggregates which cause the formation of the detected ASR gel. This method is not applied practically yet as further studies still need to be done on its application.

4.9 Discussion and Conclusion

Different AAR testing methods were discussed in this chapter. The testing methods each have different applications for which they are suitable. The tests mentioned above include tests performed to assess whether certain aggregates are susceptible to AAR; tests to assess the performance of specific concrete mixes and thus determine if they are susceptible to AAR, and also tests performed to assess the occurrence and extent of AAR in existing concrete structures. Table 19 is a summary of the different testing methods, outline the suitability of their application as well as the strength of each method in terms of ease of application, cost and practicality. The summary also describes the challenges that may be faced when considering to implement each avoidance measure.

Table 19: Summary of Strengths and Challenges of AAR Testing Methods

	Testing Method	Suitability of Application	Strengths	Challenges
1	Integrated Schemes According to RILEM and North American Guidelines			
	• Petrographic examination	Used as the first step in AAR potential assessment of aggregates	Can identify potentially reactive aggregate types, and can also identify aggregates from carbonate rocks which are produced in quarrying activities	It depends on the assessor having prior knowledge of the types of silica that are reactive in that specific region, and also depends on the petrographer being skilled and experienced.
	• Accelerated mortar-bar test for screening (AMBT)	Used to evaluate the potential of alkali-silica reactivity of aggregates in concrete.	The main advantage of employing this method is that results can be obtained within a few weeks (2 weeks) or even a few days in the	• Higher level of acceleration will tend to produce unreliable results. • Many aggregates that have given wrong or misleading results when using this test.

			extremely accelerated methods.	• Unable to detect some argillaceous dolomitic limestone aggregates in Canada that have a history of being associated with alkali-carbonate reactivity.
	• Concrete prism tests	Performed over a longer period (one year) for more accurate results, although accelerated version takes 15 weeks.	• Most likely to give an accurate reflection of the behaviour in real concrete structures • More effective in the identification of slowly reacting aggregates	•Lengthy assessment time
2	Concrete Performance Tests	Performance tests of the actual concrete mix that is going to be used in the construction of the concrete structure	Since the actual concrete mix is tested, more accurate decisions can be made with regards to the concrete to be used in the construction.	Inability to achieve adequate acceleration to provide answers in a useful timescale while at the same time giving a reliable assessment of their performance
3	AFNOR P 18-588 Microbar Test	Similar to AMBT	Similar to AMBT	Similar to AMBT
4	Concrete Imaging	Suitable to assess AAR damage in existing structures	•Testing is non-destructive and can therefore be applied without interrupting operations of infrastructure •Less costly and less time-consuming than visual inspections and can therefore	Requires skill and experience to perform assessment

			performed more frequently. • Quantitative evaluations, such as assessing the total amount of surface damage.	
5	Stiffness Damage Test (SDT) and Damage Rating Index (DRI)	Suitable to assess AAR damage in existing structures	Both the SDT and DRI are able to quite accurately distinguish different levels of ASR expansion in the concrete	Not much has been reported about the practical application
6	Multi-Physics Approach (Non-Linear Acoustic Measurements and Microwave Materials Characterisation Measurements)	To assess ASR within concrete early (when ASR gel is formed but before cracking occurs). It connects gel formation to damage and expansion.	When gel is detected within the structure and ASR is expected in that particular structure, plans can be made so that the deleterious aggregates used in the particular structure are not used in other structures, or if they must be used, appropriate avoidance measures can be employed	More research should be done on this method in order for it to be practical to implement.

5. General Discussion and Conclusions

The aim of this book was to discuss types and mechanisms of AAR, and to highlight case studies of AAR incidences around the world. Commonalities and differences between the different incidences were highlighted. This book further aimed to provide a comprehensive compilation and analysis of various AAR avoidance measures as well as AAR tests that are performed worldwide and a critical analysis was done on the measures and tests that were be discussed. Overall, due to the fact that there are various AAR avoidance measures and AAR tests performed worldwide, it was deemed necessary to have a comprehensive collection of avoidance measures and testing methods, as well as thorough critical assessments of all these, that can be used by engineers and other professionals when doing structural design of concrete structures or when they are involved in the construction, quality control or condition monitoring and assessment of concrete structures.

AAR has been identified as a deterioration mechanism which is responsible for the damage of concrete structures all over the world. It was recognized that AAR has been reported in many countries around the world where it where it has had negative impacts on the serviceability of concrete structures; the structural integrity and performance of structures; as well as the cost of infrastructure management. The three main types of AAR are Alkali-Silica Reaction (ASR), Alkali-Silicate Reaction and Alkali-Carbonate Rock Reaction (ACR). Of these three, ASR is the most common type of AAR.

Different incidences of AAR from around the world were grouped by geographic location and studied. The different areas/regions that the structures were grouped into were: The United Kingdom, Russia, North America, India, Southern and Central Africa, Australia, Nordic Europe, Asia, Mainland Europe and North Africa. The different structure groups included: bridges, water retaining structures, power stations, foundations and bases, sewage works infrastructure, railway infrastructure, tunnels, concrete pavements and car parks. The most common indicators of AAR across all infrastructure in all regions were random and map cracking and gel deposits/exudence. Although most literature indicates that AAR damage only starts to manifest after a very long time, such as 20 to 50 years, the structures that were studied demonstrated that the time for damage to appear has been as short as 2 years after construction. Water retaining structures across the regions exhibited unique damage such as displaced and rotated dam walls and they also displayed a wide range of time when damage took place, ranging between 2 years and close to 30 years.

The United Kingdom, the most commonly used aggregates which caused AAR were limestone and aggregates containing chert. In Russia was the aggregates which cause AAR were those containing the aggregates flint, sandstone and andesites. In North America, the AAR susceptible aggregates used were andesitic aggregate, greywacke and aggregates which contained limestone respectively. North America was also the region where the most significant damage to infrastructure was reported. In India, the AAR susceptible aggregates were granite, diorites and quartzite. In Southern Africa, the two very common aggregate types observed were the Witwatersrand Supergroup quartzites, greywacke rocks of the Malmesbury Group as well as granite and gneiss. In Australia, quartz was the aggregate which promoted AAR. In Nordic Europe, the AAR damaged structures contained flint. In this part of the world, it was also observed that AAR was aggravated by de-icing salts and the AAR manifested as pop-outs. Caution should therefore be practiced when operating in this region so that the concrete design caters for the effects of de-icing salts. In Asia, granite and quartz was used in the AAR damaged structures. In mainland Europe, the aggregates which caused AAR were: sandstone, limestone, siliceous limestone slate, carbonate aggregate, quartzite and gneiss. In North Africa, the AAR damaged structures contained siliceous rock. North Africa was also where the shortest period from construction to AAR damage was reported.

From the above, it can be seen that there are commonalities in the types of aggregates used in certain regions and this information can be used to make decisions about the concrete material design. For unknown aggregates, or where historical data is unavailable, it is important to test the AAR potential of such aggregates before using them.

This book further discussed AAR avoidance measures that may be employed to prevent AAR in concrete structure. These measure were also then critically analysed. The measures were:
- The use of non-reactive aggregates
- Limiting alkali content of concrete
- The use of lithium and sodium compounds
- Use of fine lightweight aggregates (FLWAs)
- The use of supplementary cementitious materials
- The use of steel fibres
- Managing environmental exposure

The most commonly employed avoidance measure was found to be the use of supplementary cementitious materials and the use of lithium compounds in the form of chemical admixtures which are manufactured by various construction materials manufacturers. Both these methods have been proven to reduce the occurrence of AAR when applied in the correct quantities and ratios. The use of fine lightweight aggregates (FLWAs) and the use of steel fibres are not very widely practiced. Although using FLWAs has been proven by experimentation to have great potential, it has not been practically applied and more research on its practical applications still needs to be conducted. Steel fibres are also not as widely used mainly due to the fact that very little research was dedicated to it and other avoidance methods are preferred to it. Using non-reactive aggregates is a straightforward method of avoiding AAR, however, unless such aggregates are locally found in the area of construction, it becomes very challenging to import non-reactive aggregate. Aggregates make up the bulk of the concrete volume and importing non-reactive aggregate would result in high transportation costs and may therefore not be economically feasible. If, however, non-reactive aggregates can be locally sourced, this would be the best avoidance measure to employ. Limiting the alkali content is a viable method if it is possible to have access to low alkali cement and also if it is possible to limit the alkali that is supplied by other alkali sources such as supplementary cementitious materials, some aggregates, deicing salts, some chemical admixtures and seawater. Managing environmental exposure is another avoidance measure which was discussed and would require the implementation of measures such as drainage in order to remove moisture that the structure is exposed to. The choice of which measure to employ depends on which methods are most practical, most economical and most sustainable.

Furthermore, this book discussed test methods employed to assess AAR. These tests were divided into the following groups:

- Tests performed to assess whether certain aggregates are susceptible to AAR, namely the petrographic examination, accelerated mortar-bar test for screening and concrete prism tests, which all form part of the integrated schemes according to RILEM and North American guidelines;
- Tests to assess the performance of specific concrete mixes and determine if they are susceptible to AAR, namely the concrete performance test in which the exact concrete mix to be used in construction is used to cast the test specimens;
- Tests performed to assess the occurrence and extent of AAR in existing concrete structures, namely concrete imaging using the GLCM texture and AAN classification; using the Stiffness Damage Test (SDT) and the Damage Rating Index (DRI) tools and the multi- physics approach method.

The main tests employed are those that are outlined in the integrated schemes according to RILEM and North America. These schemes provide systematic and methodical guidelines to assess AAR. These schemes start off with petrographic examination (which is used to identify the siliceous phases in a specific aggregate), followed by accelerated mortar bar test (which is an accelerated screening test used to evaluate the alkali-silica reactivity of aggregates in concrete) and then followed by the concrete prism tests (which are longer-term tests that give more representative results). There are also concrete performance tests that may be employed with the concrete mix design to be used in the construction and therefore give a good indication of how the concrete is expected to perform in operation. Additionally, there are also other methods that can be employed to test the occurrence of AAR in existing concrete structures. Concrete imaging, the Stiffness Damage Test (SDT) and the Damage Rating Index (DRI) are employed for this purpose and these methods are very useful to compiling the maintenance, repair and rehabilitation strategies and plans for existing concrete structures. Another method that was studied was the multi- physics approach which can be used to assess AAR in existing structures by drawing a connection between gel formation, concrete damage and concrete expansion. This method is not applied practically yet and further studies still need to be done on it. Chapter 3 provided a summary of each avoidance measure that was discussed, outlining the strength of the measure and the challenges associated with each method.

Overall, there are a number of tests in the world that can be performed on aggregates, on concrete mixes and well as on existing structures that will give sufficient information to assist in avoiding using materials susceptible to AAR, as well as to assist with the management and maintenance of concrete structures where AAR had been identified. Chapter 4 provided a summary of each testing method that was discussed, outlining the suitability of its application, as well as the strength of the test and the challenges associated with implementing the test.

6. Recommendations and Further Work

This book identified some recommendations and further work that may be conducted with regards to AAR-damaged infrastructure worldwide; AAR avoidance measures and AAR tests worldwide. These are:

- There is a severe lack of information about the actual cost of the management of infrastructure damaged by AAR. This can most likely be partially attributed to the fact that AAR damage usually occurs together with damage caused by other deterioration mechanisms, which makes it challenging to isolate the damage caused by AAR. It is however important for an effort to be made to quantify the financial cost of AAR, so that this can be linked to efforts dedicated to the avoidance measures as well the testing methods employed. Future work in this regard is recommended.

- The use of steel fibres as an effective AAR avoidance measure has proven to have a lot of potential. Using fibres in concrete is practiced for other purposes, and could therefore be extended to prevent AAR. Performing further studies regarding practicality of using this method is recommended.

- Concrete performance tests using the actual concrete mix for construction are currently being considered by RILEM. Once these tests are finalized, they will be very useful because they will be able to provide a very accurate picture of how concrete is expected to behave. This future work is anticipated and encouraged by the author.

- The SDI and DRI methods have the potential to distinguish the different levels of ASR. Further studies into this work is recommended as they can be very useful in the assessment, management, repair and rehabilitation of structures affected by AAR.

- The multi-physics method needs further work as it has the potential to provide very useful information regarding the possibility of ASR damage in the future, even before an ASR-affected structure starts cracking. This would be in managing the future of structures to be constructed with the same materials as those used in the damaged structure.

References

Alexander, M., & Beushausen, H. (2009). Deformation and volume change of hardened concrete. In G. Owens (Ed.), *Fulton's Concrete Technology* (9th ed., pp. 111–154). Cape Town: Cement & Concrete Institute.

Alexander, M., & Blight, G. E. (2011). *Alkali-Aggregate Reaction and Structural Damage to Concrete* (Vol. 4). London: CRC Press/Balkema.

Alexander, M., & Blight, G. E. (2017). Southern and Central Africa. In I. Sims & A. Poole (Eds.), *Alkali-Aggregate Reaction in Concrete - A World Review* (pp. 510–538). London: CRC Press/Balkema.

Alexander, M. G. (2019). Alkali-Aggregate Reaction. In S. Mindess (Ed.), *Developments in the formulation and reinforcement of concrete* (2nd ed., pp. 87–113). Cape Town: Woodhead Publishing.

ASTM. (2003). Standard Guide for Petrographic Examination of Aggregates for Concrete. *ASTM International, 04*, 1–9. https://doi.org/10.1520/C0295

ASTM. (2016). Standard Test Method for Potential Alkali-Silica Reactivity of Aggregates (Chemical Method) (Withdrawn 2016). https://doi.org/10.1520/C0289-07

ASTM. (2019). *Standard Test Method for Potential Alkali Reactivity of Cement- Aggregate Combinations (Mortar-Bar Method) (Withdrawn 2018)*. 1–2. https://doi.org/10.1520/C0227-10

Beushausen, H. (2011a). *Namwater Assessment of concrete structures (water resevoirs) Reservoir: Collector 2*. Cape Town.

Beushausen, H. (2011b). *Namwater Assessment of concrete structures (water resevoirs) Reservoir: Mile 7 (Round Reservoir)*. Cape Town.

Beushausen, H. (2011c). *Namwater Assessment of concrete structures (water resevoirs) Reservoir: Mile 7 (Square Reservoir)*. Cape Town.

Beushausen, H. (2011d). *Namwater Assessment of concrete structures (water resevoirs) Reservoir: Rooibank*. Cape Town.

Blight, G. E., & Alexander, M. G. (2011). Alkali-Aggregate Reaction and Structural Damage to Concrete. In *Alkali-Aggregate Reaction and Structural Damage to Concrete*. Leiden: CRC Press.

BS. (1994). BS 812-104:1994 Testing aggregates. Method for qualitative and quantitative petrographic examination of aggregates. *British Standard Intitution*.

Clark, L. A. (1989). *Critical review of the structural implications of the alkali-silica reaction in concrete*. Crowthorne: Berkshire: Transport and Road Research Laboratory.

Crisp, T., Waldron, P., & Wood, J. (1993). *Development of a non destructive test to quantify damage in deteriorated concrete. Magazine of Concrete Research. 45*(165), 247–256.

de Carvalho, M. R. P., Fairbairn, E. de M. R., Filho, R. D. T., Cordeiro, G. C., & Hasparyk, N. P. (2010). Influence of steel fibers on the development of alkali-aggregate reaction. *Cement and Concrete Research, 40*(4), 598–604.

Falikman, V., & Rozentahl, N. (2017). Russian Federation. In S. Ian & A. Poole (Eds.), *Alkali-Aggregate Reaction in Concrete - A World Review* (pp. 433–456). London: CRC Press/Balkema.

Fernandes, I., Andic-Cakir, O., Giebson, C., & Seyfarth, K. (2017). Mainland Europe,Turkey and Cyprus. In I. Sims & A. Poole (Eds.), *Alkali-Aggregate Reaction in Concrete - A World Review* (pp. 312–432). London: CRC Press/Balkema.

Godart, B., & de Rooij, M. (2013). Guide to Diagnosis and Appraisal of AAR Damage to Concrete in Structures. In W. J. G. M. Godart B., de Rooij M. (Ed.), *RILEM State-of-the-Art Reports* (Vol. 12). New York: Springer.

Hasparyk, N. P., Farias, L. A., Andrade, M. A. S., & Bittencourt, R. M. (2004). *Study of the Influence of Amorphous Rice Husk-Ash on Concrete Properties*. (November), 751–760.

Hooton, R. D., Rogers, C., MacDonald, C. A., & Ramlochan, T. (2013). Twenty-year field evaluation of alkali-silica reaction mitigation. *ACI Materials Journal, 110*(5), 539–548.

Hooton, R., & Rogers, C. (2003). Development of the NBRI rapid mortar bar test leading to its use in North America. *Construction and Building Materials, 7*(3), 145–148.

Kabir, S. (2006). *Analysis of surface-damage through concrete imaging*. (September).

Kay, T., Poole, A. B., & Sims, I. (2017). Middle East & North Africa. In I. Sims & A. Poole (Eds.), *Alkali-Aggregate Reaction in Concrete - A World Review* (pp. 729–756). London: CRC Press/Balkema.

Leemann, A., Thalmann, C., & Studer, W. (2005). Alkali-aggregate reaction in Swiss tunnels. *Materials and Structures/Materiaux et Constructions, 38*(277), 381–386.

Li, C., Thomas, M. D. A., & Ideker, J. H. (2018). A mechanistic study on mitigation of alkali-silica reaction by fine lightweight aggregates. *Cement and Concrete Research, 104*(October 2017), 13–24.

Lindgard, J., Grelk, B., Wigum, B., Tragardh, J., Appelqvist, K., Ferreira, M., & Leivo, M. (2017). Nordic Europe. In S. Ian & A. Poole (Eds.), *Alkali-Aggregate Reaction in Concrete - A World Review* (pp. 277–320). CRC Press/Balkema.

Mackechnie, J. (2021). Alkali Silica Reaction. In G. Owens (Ed.), *Fulton's Concrete Technology* (10th ed.). Midrand: The Concrete Institute.

Mullick, A. K. (2017). Indian Sub-Continent. In I. Sims & A. Poole (Eds.), *Alkali-Aggregate Reaction in Concrete - A World Review* (pp. 603–727). London: CRC Press/Balkema.

Nixon, P., & Fournier, B. (2017). Assessment, Testing and Specification. In I. Sims & A. B. Poole (Eds.), *Alkali-Aggregate Reaction in Concrete - A World Review* (pp. 33–61). London: CRC Press/Balkema.

Nixon, P. J., & Sims, I. (2016). *RILEM Recommendations for the Prevention of Damage by Alkali- Aggregate Reactions in New Concrete Structures*. London: Springer.

Oberholster, B. (2009). Alkali Silica Reaction. In G. Owens (Ed.), *Fulton's Concrete Technology* (9th ed., pp. 189–218). Midrand: Cement and Concrete Institute.

Prasetia, I., & Torii, K. (2013). *the Mitigation Effects of Lithium and Sodium Compounds on Alkali-Silica Reaction of Concretes*. 1–6.

Rashidi, M., Knapp, M. C. L., Hashemi, A., Kim, J. Y., Donnell, K. M., Zoughi, R., … Kurtis, K. E. (2016). Detecting alkali-silica reaction: A multi-physics approach. *Cement and Concrete Composites, 73*, 123–135.

SABS. (2013). *SANS 1083 : 2013 South African National Standard Aggregates from natural sources — Aggregates for concrete*.

Sanchez, L., Fournier, B., Jolin, M., & Bastien, J. (2016). *Assessment of Damage Due to Alkali-Silica Reaction Through Microscopic and Mechanical Tools*. 236–243.

Shayan, A., & Freitag, S. (2017). Australia and New Zealand. In I. Sims & A. Poole (Eds.), *Alkali-Aggregate Reaction in Concrete - A World Review* (pp. 570–681). London: CRC Press/Balkema.

Sika. (2014). *Sika ® Control ASR Admixture For Alkali-Silica Reaction Mitigation Sika ® Control ASR*.

Silva, A. S., Riberio, A. B., Jalali, S., & Divet, L. (2006). *Use of Fly Ash and Metakaolin for*

the Prevention of Alkali-Silica Reaction and Delayed Ettringite Formation in Concrete.
Sims, I. (2017). United Kingdom and Ireland. In S. Ian & A. Poole (Eds.), *Alkali-Aggregate Reaction in Concrete - A World Review* (pp. 167–276). London: CRC Press/Balkema.
Sims, I., & Poole, A. (2017). *Alkali-Aggregate Reaction in Concrete - A World Review* (I. Sims & A. Poole, Eds.). London: CRC Press/Balkema.
Stefan, Gó. (2012). Alkali - carbonate reaction of aggregates. *Gospodarka Surowcami Mineralnymi / Mineral Resources Management, 28*(1), 45–62.
Thomas, M., Folliard, K. J., & Ideker, J. H. (2017). North America (USA and Canada). In I. Sims & A. Poole (Eds.), *Alkali-Aggregate Reaction in Concrete - A World Review* (pp. 467–491). London: CRC Press/Balkema.
Thomas, M., Hooton, R. D., & Folliard, K. J. (2017). Preventionof Alkali-Silica Reaction. In I. Sims & A. Poole (Eds.), *Alkali-Aggregate Reaction in Concrete - A World Review* (pp. 89–118). London: CRC Press/Balkema.
Yamada, K., & Miyagawa, T. (2017). Japan, China and South-East Asia. In I. Sims & A. Poole (Eds.), *Alkali-Aggregate Reaction in Concrete - A World Review* (pp. 540–568). London: CRC Press/Balkema.

APPENDICES

Appendix 1: The Use of Lithium and Sodium Compounds - Japan

A research was conducted in Japan to investigate the diffusivity of externally supplied lithium and sodium compounds into cementitious material. The purpose of this research was to investigate the possibility of using a more economical and effective compound for the mitigation of ASR in concrete (Prasetia & Torii, 2013). The diffusivity was tested using the diffusion cell test. Mortar bar tests were then employed to measure the expansion of test specimens and thus test the mitigation effects of lithium and sodium compounds on ASR in concrete (Prasetia & Torii, 2013).

Experimental Methods

Diffusion Cell Test

The cement pastes used for the experiments were prepared with Ordinary Portland Cement (OPC) with a water/cement ratio of 0.4, referred to as OPC-0.4. Two lithium and two sodium electrolytes were used in the experiment, namely (Prasetia & Torii, 2013) :

- $LiNO_3$, at a concentration of 1 mol/L
- Li_2SiO_3, at a concentration of 0.5 mol/L
- $NaNO_3$, at a concentration of 1 mol/L
- Na_2SiO_3, at a concentration of 0.5 mol/L

The diffusion cell test was carried out using the diffusion cell unit shown in Figure 41. Cement paste rings of 30 mm diameter and 5mm thickness were cast and then allowed to cure in Ca $(OH)_2$ saturated solution for 7, 28 and 91 days at a room temperature of 20 °C and 60% relative humidity. The diffusion cells in Figure 41 were composed of 100ml of the electrolyte sources of the lithium of sodium on the right hand said, referred to as the tracer cell, and 100ml of deionised water on the left hand side, referred to as the measurement cell (Prasetia & Torii, 2013). The concentration of the ions was measured using a PIA-100 ion analyser and the results of the concentration are shown in Figure 42 and Figure 43. The following may be summarised about the results (Prasetia & Torii, 2013):

- In the lithium solutions, there was a steep rise in the diffusion gradient of Li+ ions from the $LiNO_3$ solution at 28 days of curing and for the Li_2SiO_3 solution at 7 of days curing
- In the lithium solutions, the $LiNO_3$ solution at 7 and 91 days of curing and for the Li_2SiO_3 solution at 28 and 91 days of curing, increase in diffusion gradient was less evident
- In the sodium solutions, the diffusion concentration of Na+ ions was large at 28 days of curing but reduced at 91 days of curing, showing that $NaNO_3$ is absorbed well into the hydration products of the paste

-In general, the concentration of sodium solutions was 10 times greater than the lithium solutions.

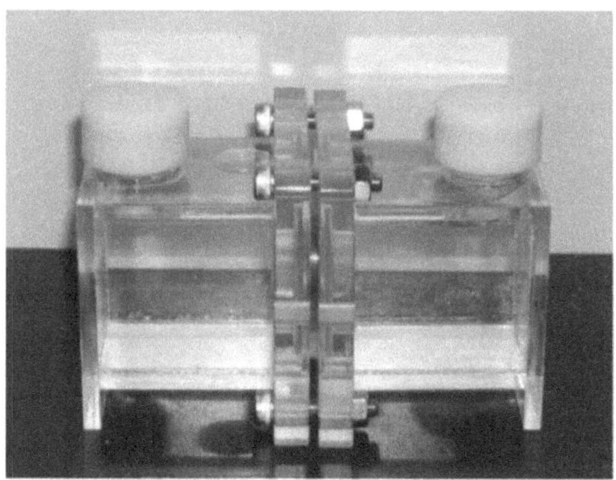

Figure 41:Diffusion cell unit (Prasetia & Torii, 2013)

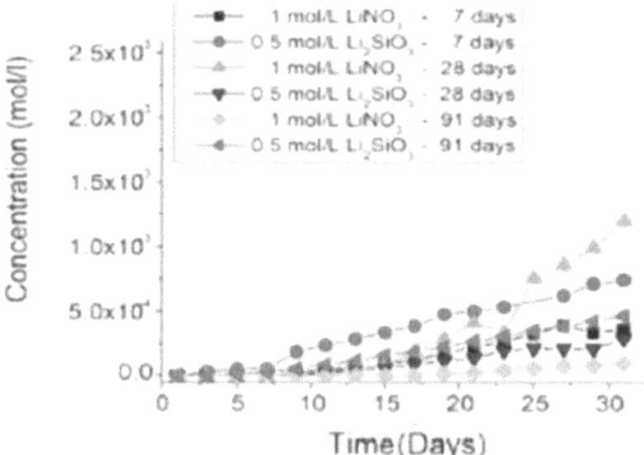

Figure 42: Diffusion profiles of Li+ from LiNO3 and Li2SiO3 solution (curing time: 7, 28 and 91 days) (Prasetia & Torii, 2013)

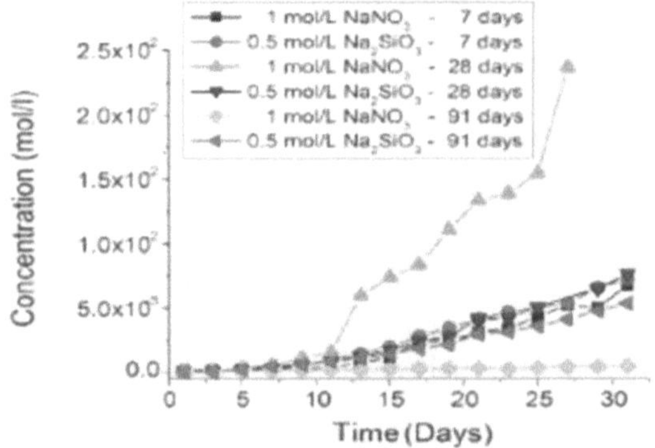

Figure 43: Diffusion profiles of Na+ from NaNO3 and Na2SiO3 solution (curing time: 7, 28 and 91 days) (Prasetia & Torii, 2013)

Accelerated Mortar Bar Test

The accelerated mortar test in accordance with ASTM C1260 was employed for the test specimens which were immersed in solutions with high and low lithium and sodium concentrations (Prasetia & Torii, 2013). The following materials were used for the mortar bar test (Prasetia & Torii, 2013):

- The reactive aggregate was crushed calcined flint from industrial raw materials in England, with grain size range of 0.6 mm to 2.5mm
- The non-reactive aggregate was crushed limestone sand with a particle size range of 0.15mm to 5mm
- Ratio of calcined flint to crushed limestone was 25%:75%
- Ordinary Portland cement with density of 3.16 g/cm3, Blaine specific area of 3300 cm2/g and alkali content of 0.42%

For the mortar bar test, the 25 x 25 x 285mm specimens were immersed as follows, in the specified concentration categories (low or high) (Prasetia & Torii, 2013):

- 0.5 mol/L concentrated Li2SiO3 (for low concentration)
- 0.5 mol/L concentrated Na2SiO3 (for low concentration)
- 2.5 mol/L concentrated Li2SiO3 (for high concentration)
- 2.5 mol/L concentrated Na2SiO3 (for high concentration)
- 1 mol/L concentrated NaNO3 (for low concentration)

Results

Figure 44 (Prasetia & Torii, 2013) shows the expansion of the immersed specimens.

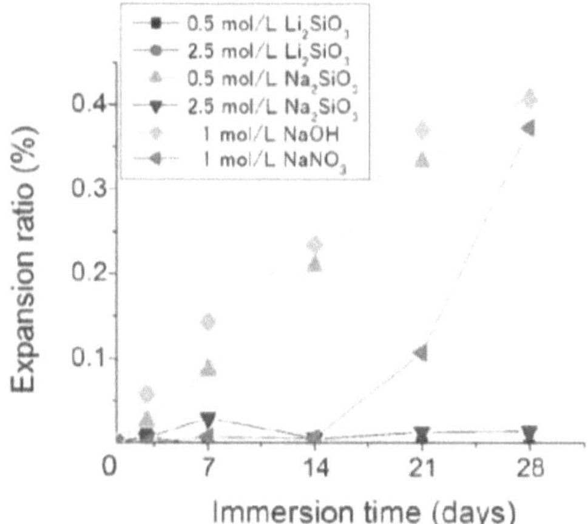

Figure 44: Expansion ratio of accelerated mortar bars immersed in various lithium and sodium compounds in accordance with ASTM C1260 (Prasetia & Torii, 2013)

The results may be summarised as follows (Prasetia & Torii, 2013):

- Specimens which were immersed in Li_2SiO_3 exhibited no expansion, regardless of the concentrations. This may be taken to mean that no ASR took place in these specimens.

- Specimens that were immersed in low concentration (0.5 mol/L) Na_2SiO_3 solution exceeded 0.2% expansion at 14 days and exceeded 0.4% expansion at 28 days

- Specimens that were immersed in low concentration (1 mol/L) NaNO3 had no expansion up to 14 days but had an expansion exceeded 0.3% at 28 days

- Specimens that were immersed in high concentration (2.5 mol/L) Na_2SiO_3 solution had a very small early expansion, which disappeared. Unlike with the low concentration Na_2SiO_3, which was in a liquid state at 80 °C, the high concentration Na_2SiO_3 solution had a high viscosity and was in a paste-like state until 14 days.

- A likely explanation for the Na_2SiO_3 solution behavior is that it was difficult for the highly concentrated, water-glass solution to penetrate the test specimens while in the presence of the highly reactive flint aggregates. The hydrolysis reaction between the sodium compound and the water produces NaOH, which increases the alkalinity of the pore solution. It also increases the OH- in the pore solution, starting the ASR. Increasing the concentration of Na_2SiO_3 from an external source however, triggers the reaction between Na_2SiO_3 and $Ca (OH)_2$ and it surpasses the pace of the alkali silica reaction. An overwhelming amount of C-S-H gel is produced and engulfs the ASR gel produced previously, thus suppressing the expansion. This activity is different from that of the lithium compounds due to the fact that highly reactive lithium stops ASR from occurring in the first place

It can be concluded from the experiments conducted that the diffusion coefficients of Na+ ions were 10 times greater than those of Li+, which indicates that lithium compounds are better absorbed into the hydration compounds and they react better with them, compared to Na+ ions. Additionally, no expansion of the mortar bar is observed regardless of the Li_2SiO_3, which shows that there was no ASR took place and therefore Li_2SiO_3 can be used to mitigate expansion due to ASR in concrete (Prasetia & Torii, 2013). Na_2SiO_3 may also be used to mitigate ASR-induced expansion in concrete when applied at high concentration. These measures may be used as cost effective ASR avoidance/mitigation measures in concrete.

Appendix 2: Use of fine lightweight aggregates (FLWAs)

A research was conducted to determine whether fine lightweight aggregates (FLWAs) can be used to mitigate alkali silica reaction in concrete.

Materials

In the research, three FLWAs were investigated for their mitigation impacts, namely: expanded slate, shale and clay. FLWAs have a higher absorption capacity and a lower specific gravity compared to normal fine aggregate. These are shown in Table 20. The XRF results for the three FLWAs are shown in Table 21.

Table 20: Material Properties of the Investigated FLWAs (Li et al., 2018)

MATERIAL PROPERTIES OF THE INVESTIGATED FLWAs		
FLWAs	Absorption Capacity (%)	Specific Gravity
Expanded Slate	10.27	1.74
Expanded Shale	23.34	1.52
Expanded Clay	30.75	1.07

Table 21: XRF Data of the Investigated FLWAs (Li et al., 2018)

XRF DATA OF THE INVESTIGATED FLWAs (%)										
	SiO_2	Al_2O_3	Fe_2O_3	CaO	MgO	SO_3	Na_2O	K_2O	TiO_2	P_2O_5	Total Oxides
Expanded Slate	61.35	18.53	7.82	2.76	2.11	0.64	2.70	3.13	0.80	0.16	100.0
Expanded Shale	61.83	19.61	7.72	3.33	2.75	0.64	0.52	2.87	0.61	0.12	100.0
Expanded Clay	68.04	15.77	5.54	2.70	2.21	0.74	1.12	3.06	0.61	0.20	100.0

In the research, two fine aggregates and two coarse aggregates with varying reactivity were also used. The reactivity of these aggregates was classified in accordance with ASTM C1778. These details of the aggregates are shown in *Table 22*.

Table 22: Aggregates Used in the Study (Li et al., 2018)

COARSE AND FINE AGGREGATES USED IN THE STUDY (IN ACCORDANCE WITH ASTM C1778)				
	ID	Type	Reactivity	Source
Coarse Aggregate	CA1	Crushed limestone	Non-reactive	Washington, USA
	CA2	Spratt III	Highly reactive	Ontario, Canada
Fine Aggregate	FA1	Siliceous river sand	Very highly reactive	Oregon, USA
	FA2	Crushed limestone	Non-reactive	Texas, USA

The cement used in the research was a high-alkali Portland cement Type I as per ASTM C150 classification. The chemical composition of the cement is shown in Table 23.

Table 23: Chemical Composition in Cement (Li et al., 2018)

MAJOR CHEMICAL COMPOSITION IN CEMENT (%)									
SiO_2	Al_2O_3	Fe_2O_3	CaO	MgO	Na_2O	K_2O	Na_2O_{eq}	SO_3	LOI
19.61	4.38	2.76	62.21	2.72	0.28	0.84	0.83	3.76	2.60

Methodology

The accelerated mortar bar test (AMBT) in accordance with ASTM C1260 was employed to test the reactivity of the aggregates. The AMBT evaluates the aggregate expansion after 14 days of exposure to 1N NaOH at 80°C. A reference mix was made with a very highly reactive siliceous sand (FA1). The FLWAs were prepared either pre-wetted or oven dried and they were used to replace FA1 at 25%, 50% and 100% by volume separately, making up a total of 17 mixes with different replacement values of FWLAs (Li et al., 2018) . The 17 mixes are shown in Table 24.

Table 24: Mortar Mixes used in Experiment (AMBT) (Li et al., 2018)

MIXTURE MATRIX FOR AMBT (17 MIXTURES INCLUDING ONE REFERENCE)					
Aggregate	FLWA	Treatment	Replacement Level (%)		
FA1	N/A	Oven-dried	-		
FA1	Expanded slate	FA1: oven-dried	25	50	100
	Expanded shale	FLWAs: pre-wetted	25	50	100
	Expanded clay		25	50	100
FA1	Expanded slate	FA1: oven-dried	25	50	100
	Expanded shale	FLWAs: oven-dried	25	50	-
	Expanded clay		25	50	-

The concrete prism test (CPT) was also employed to evaluate the aggregate reactivity over a one year period or mitigation efficacy over a two-year period. The test prisms were stored at 38°C and 100% relative humidity. The CPT takes longer than the AMBT but gives a better indication of field performance. Table 25 shows two reference mixes which were evaluated. In the CPT, two reference mixes were evaluated, namely BN and SP. In the BN mixture, a highly reactive fine aggregate and a non-reactive coarse aggregate were used, while in the SP mixture, a highly reactive coarse aggregate an a non-reactive fine aggregate were used.

Table 25: Mixture Matrix for CPT (Li et al., 2018)

MIXTURE MATRIX FOR CPT					
	Coarse Aggregate		Fine Aggregate		
Group	Aggregate Type	Reactivity	Aggregate Type	Reactivity	
BN	Crushed limestone (CA1)	Non-reactive	Siliceous river sand (FA1)	Very highly reactive	
SP	Spratt III (CA2)	Highly reactive	Crushed limestone (FA2)	Non-reactive	

Results

The results for the AMBT are shown in Figure 45. The FLWAs were prepared either pre-wetted or oven dried and they were used to replace FA1 at 25%, 50% and 100% by volume. Figure 45 shows the expansions of the two pre-treatment groups (pre-wetted and oven-dried) of samples after they were exposed to 1.0 N NaOH for 14 days. The results are summarized as follows (Li et al., 2018):

- The reference sample with the highly reactive fine aggregate FA1 had an expansion of 0.66%, which significantly exceeded 0.1% which is the expansion limit.
- With the exception of the 25% pre-wetted expanded slate, all expansions were reduced by with an increase in FLWA replacement in both the pre-wetted and oven dried groups.
- In both groups, the mix with 100% expanded shale was below the expansion limit of 0.1%, indicating that expanded shale was non-reactive
- Expanded clay and expanded shale limited expansion to a better extent that expanded slate
- All mixtures with pre-wetted FLWAs with 100% replacement of FLWAs showed behaviour which was not deleteriously reactive.
- For the oven-dried mixtures, the 25% expanded slate had an expansion of 0.55%, while the 25% expanded slate of the pre-wetted mixtures had an expansion of 0.77%
- The AMBT results show that the mixtures containing the pre-wetted FLWAs had a higher expansion than those containing oven-dried FLWAs at the same quantity of replacement.

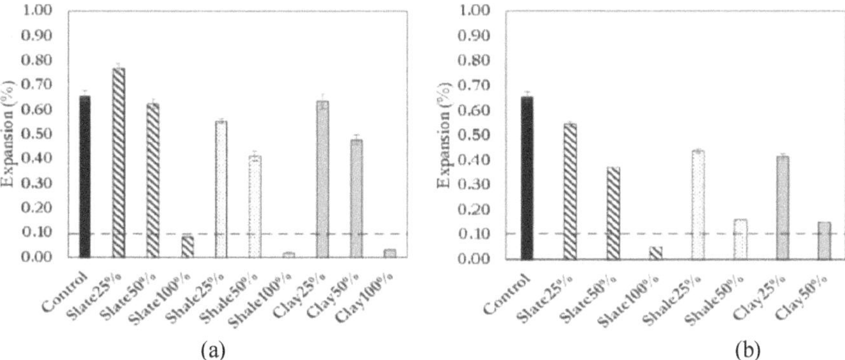

Figure 45: AMBT expansion results after 14 days. (a) Mixtures with pre-wetted FLWAs and (b) Mixtures with oven-dried FLWAs (Li et al., 2018)

The results of the CPT for two years are shown in Figure 46. In this tests, all the FLWAs were pre-wetted so that the moisture content was above absorption capacity. The results of the CPT are summarised as follows (Li et al., 2018):

- In the BN mixtures, the reference showed an expansion of 0.66%
- The 25% expanded slate showed a minimal impact on expansion
- From all three FLWAs, expanded clay was most effective in reducing expansion due to ASR
- Mixtures in which FLWAs replaced highly reactive fine aggregates showed a reduction in expansion in both the AMBT and the CPT.

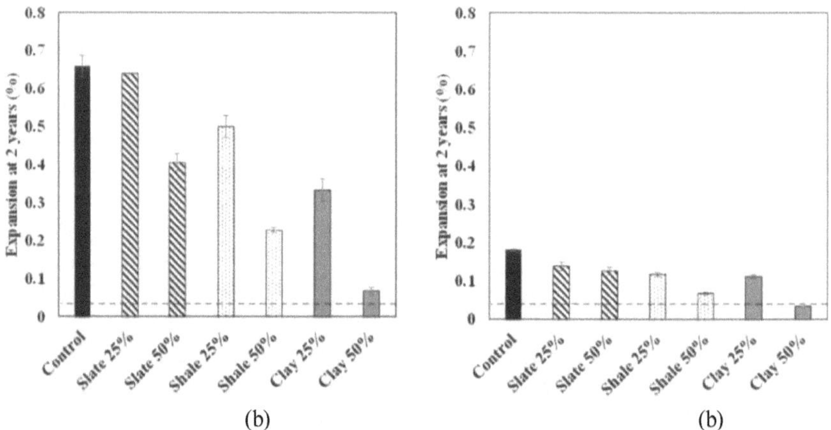

Figure 46: Expansion results of CPT at 2 years. (a) BN mixtures and (b) SP mixtures (Li et al., 2018)

From the results of this investigation, the following is concluded about the impact of FLWAs on mitigating expansion due to ASR (Li et al., 2018):

- Expanded slate, shale and clay all effectively reduce the concrete expansion due to ASR
- An increase in the replacement level of FLWAs results in less expansion in concrete.

Appendix 3: Fly Ash and Metakaolin – Portugal

A research was conducted to study the influence of supplementary cementitious materials such as fly ash and metakaolin on inhibiting the occurrence of ASR in concrete (Silva et al., 2006).

Materials

Concrete was made with high alkali Portland cement CEM I 42.5R, of which fly ash (FA) and metakaolin were used as partial replacements by volume of the cement. The concrete specimens was the cured at normal and elevated temperatures of up to 80°C after which they were stored in saturated ambient humidity or in water at ambient temperature for a period of up to 2 years (Silva et al., 2006). The chemical compositions of the cement, fly ash and metakaolin are shown in Table 26:

Table 26: Chemical composition of cement, fly ash and metakaolin used in this study (Silva et al., 2006)

Oxide (mass %)	Cement	Fly Ash	Metakaolin
SiO_2	18.81	53.22	54.66
Al_2O_3	5.15	23.20	37.98
Fe_2O_3	3.18	5.85	1.22
CaO	63.70	5.36	0.01
MgO	1.50	1.63	0.46
SO_3	2.69	1.00	0.01
K_2O	1.02	1.42	3.09
Na_2O	0.19	0.44	0.00
Na_2O_{eq}	0.86	1.37	2.03
$TiO_2 + P_2O_5$	0.34	1.87	0.49
LOI	3.18	5.16	0.94

Experimental Methods

Concrete testing was performed in accordance with RILEM AAR-3, which is equivalent to ASTM C 1293. The concrete mix was made as follows (Silva et al., 2006):
- 440 kg/m^3
- w/c = 0.45
- alkali content of 5.5 kg/m3 Na_2O_{eq} calculated on the basis of available alkalis from the cement
- FA
- MK
- Added NaOH

The fly ash partial replacement of the cement was done at 20, 40 and 60%, while the metakaolin partial replacement was done at 20% while maintaining a constant w/c ratio (Silva et al., 2006). The expansion results of these tests are shown in Figure 47:

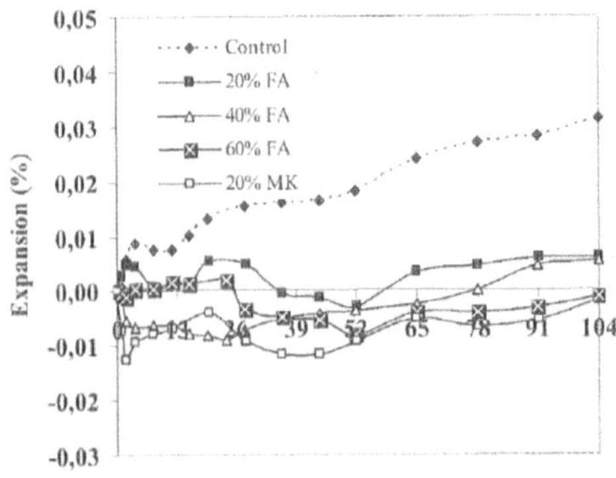

Figure 47: Expansion of concretes in accordance with the RILEM AAR-3 method (Silva et al., 2006)

Results

Figure 47 shows results that can be summarised as follows (Silva et al., 2006):

- The expansion of the control increases over time due to the reactive nature of the reactive nature of the aggregate used in the mix
- 20% replacement with FA was effective in inhibiting the expansion due to ASR
- 20% replacement with MK had an even more significant effect on inhibiting the expansion due to ASR than FA

Appendix 4: Amorphous Rice Husk-Ash – Brazil

Experiments were carried out to investigate the impact of rice husk ash (RHA) on concrete properties, among them the impact on Alkali Aggregate Reaction (AAR). The tests were carried out by partially substituting the cement with varying quantities of RHA, which were: 8%, 10%, 12%, 16% and 20% (Hasparyk et al., 2004). The RHA was used in two states, namely the natural state and the ground state.

Materials

The following materials were used for the experiment (Hasparyk et al., 2004):
- Natural sand obtained from the riverbed of the Tocantins River
- Gabbro lithological coarse aggregate with a maximum size of 25mm
- A multifunctional admixture
- CP II-E-32 Brazilian Portland Cement
- Rice husk ash from the burning of rice husk. The rice husk was used in both its natural state (where it hasn't been ground- it was used as collected from the power plant) and in its ground state (where the husk is processed in a ball mill)

Table 27 shows the chemical analysis of the RHA, where it can be seen that RHA is very siliceous material. Table 28 shows the physical characteristics and pozzolanic activity of the RHA (Hasparyk et al., 2004).

Table 27: Chemical analysis of RHA (Hasparyk et al., 2004)

Loss on Ignition	SiO_2	Al_2O_3	Fe_2O_3	CaO	MgO	Na_2O	K_2O	Na_2O_{eq}	SO_3	SiO_2+ Al_2O_3 + Fe_2O_3
12.6	79.82	0.27	3.11	0.63	0.81	0.19	1.26	1.02	0.13	83.2

Note: Na_2O_{eq} = 0.658 x K_2O + Na_2O

Table 28: Physical characteristics and pozzolanic activity of RHA (Hasparyk et al., 2004)

Properties	Natural	Ground	Threshold	
Specific gravity (g/cm3)	2.06	2.20	-	-
Average diameter by laser distribution (μm)	48	12	-	-
Residue on #325 screen (45μm)-moist method (%)	88.2	15.5	$\leq$34	NBR 12653/92
Index of pozzolanic activity with CP II-E-32 cement -28 days (%)	31.2	96.2	$\geq$75	NBR 12653/92
Index of pozzolanic activity with lime – 7 days (MPa)	1.3	12.5	$\geq$6.0	
Pozzolanic activity according to Chapelle (mgCaO/g amostra)	Not determined	706.4	$\geq$330	[1]

Note 1: Reference taken from Raverdy et al.

Test Methods

The alkali aggregate reaction test was done in accordance with ASTM C1260. The mortar specimens were manufactured with a cement: aggregate ratio of 1:2.25 by mass, and were immersed in 1N sodium hydroxide (NaOH) solution at a temperature of 80°C (Hasparyk et al.,

2004). In order to verify the potential of expansion reduction in the presence of RHA, mortar samples were cast with the reactive aggregate pyrex glass which is in the form of an artificial sand as well as Ordinary Brazilian Portland with admixture (CP II-F-32) with a high alkali content (1.16% of total Na_2O_{eq} and 0.86% of soluble Na_2O_{eq}). RHA was used to replace cement volume in contents of 0%, 8%, 10%, 12%, 16% and 20% for both natural and ground ash types (Hasparyk et al., 2004).

Results

The AAR expansion reduction potential was determined relative to the reference sample and presented in percentage. The results for both types of ash are shown in Figure 48 and Figure 49 at tests ages of 16 and 30 days.

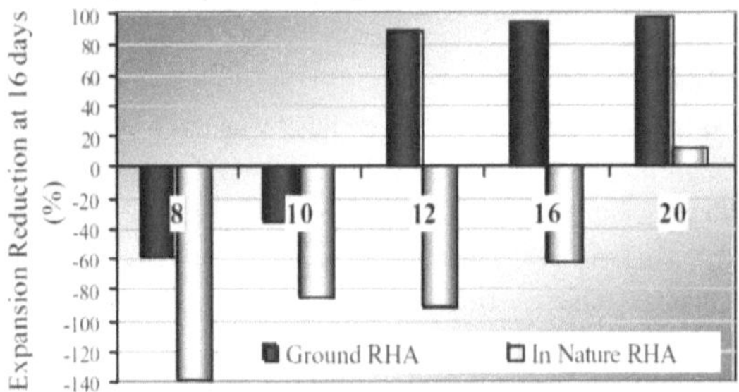

Figure 48: Reduction in expansions at 16 days of testing (Hasparyk et al., 2004)

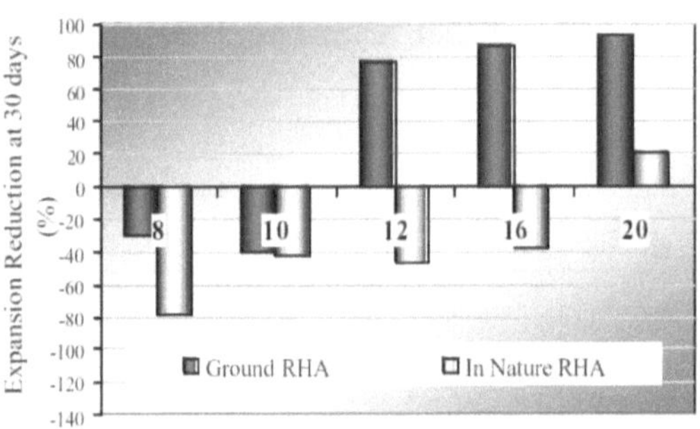

Figure 49: Reduction in expansions at 30 days of testing (Hasparyk et al., 2004)

The following is summarised from the results (Hasparyk et al., 2004):

- In the specimens with ground RHA, the ones with 12%, 16% and 20% contents showed a significantly reduced expansions below 0.1% which is the limit as prescribed by ASTM C1260. This applies to both 16 days and 30 days. Increasing the content of RHA increased the effectiveness of the expansion reduction.

- In the specimens with natural RHA, all the contents (with the exception of 20%), the presence of RHA did not reduce expansions at 16 days. Only a content of 20% RHA showed a reduction in expansion, and even that expansions were above the prescribed limits, meaning that reactive behaviour was still present.

It can therefore be seen from the results of this experiment that ground RHA, when used as a partial replacement of cement can successfully reduce expansion due to AAR. This effect can be seen from a replacement value of 12% and increases with increasing RHA content.

Appendix 5: The Use of Steel Fibres - Brazil

An experiment was conducted by the Federal University of Rio de Janeiro in Brazil in conjunction with the Brazilian hydropower industry, to test the impact of steel fibres on mortar specimens which were subjected to Alkali Aggregate Reaction (AAR) (de Carvalho et al., 2010). Two types of steel fibres were used in the experiment, and the fibre contents used were 1% and 2%. The two types of steel fibres used were of dimensions:
- 0.16 mm diameter and 6mm length, and
- 0.2 mm diameter and 13mm length.

Materials

The cement used in the experiment was a Portland cement, which was equivalent to ASTM type 1, with a high alkali content. Table 29 shows the characteristics of the cement.

Table 29: Main physiochemical properties for cement [a] (de Carvalho et al., 2010)

Density		3.12 g/cm^3
Residue on sieve 200		1.7
Residue on sieve 325		12.2
Specific surface area		$311 \text{ m}^2/\text{kg}$
Compressive strength	3 days	20.4 MPa
	7 days	25.9 MPa
	28 days	31.3 MPa
Loss on ignition		1.55
Insoluble residue		0.35
SO_3		2.44
MgO		1.16
SiO_2		20.75
Fe_2O_3		2.56
Al_2O_3		4.52
CaO		64.65
Free lime		3.0
Total alkali	Na_2O	0.77
	K_2O	0.84
	Alkali equivalent [b]	1.32
Soluble alkali	Na_2O	0.34
	K_2O	0.75
	Alkali equivalent [b]	0.83
$CaSO4$		4.15

[a] Units in % by mass where not specified
[b] Na2Oeq = Na2O + 0.658x K2O

The reactive aggregates used were diabasic rocks. According to petrography analysis performed on the rocks, it contained a fine grained matrix which suggests the volcanic glass indicating reactive potentiality. According to the ASTM C1260 test method, this rock was considered deleterious (de Carvalho et al., 2010). The grain size distribution of the aggregate used is shown in Table 30:

Table 30: Grain size distribution (de Carvalho et al., 2010)

Sieve	% by mass
4.8 mm to 2.4 mm	10
2.4 mm to 1.2 mm	25
1.2 mm to 0.6 mm	25
0.6 mm to 0.3 mm	25
0.3 mm to 0.15 mm	15

The two types of steel fibres used were:
- 0.16 mm diameter and 6mm length, with an aspect ratio (L/D) of 37.5, and
- 0.2 mm diameter and 13mm length with an aspect ratio (L/D) of 65.

Experimental Methods

The test was conducted by casting five different mortars with Portland cement, the diabasic aggregate, as well as varying contents of steel fibre volumes. The contents of the steel fibres were 0%, 1%and 2% for the two fibre types. The five mortar mixes were: 0%SF, 1%SF6, 2%SF6, 1%SF13 and 2%SF13, where the first number denotes the fibre volume content (0%, 1% or 2%), the SF denote 'steel fibre', and the last number denotes the length of the fibre length (either 6mm or 13mm). The mixture 0%SF was the reference mixture (de Carvalho et al., 2010).

The test was done in accordance with ASTM C1260. Some modifications were made to the test in order to meet certain condition of the fibre-reinforced material. One such modification was that instead of the standard specimen size of 25 x 252 x 285 mm, a size of 75 x 75 285 mm was used. The three specimens were cured for 28 days in a high-humidity chamber as per the standard (21 ± 1°C and 100% relative humidity). They were then immersed in a water bath with a temperature up to 60°C for 4 hours. They were then transferred to a bath containing sodium hydroxide 1N solution at a temperature of 60 ± 2°C for 60 days and length measurements of the mortar samples were taken in accordance with ASTM C490 (de Carvalho et al., 2010).

Results

The expansion results of the different mortar specimens are shown in Figure 50. The time (days) in the figure indicated the amount of time the specimens were in the NaOH solution, which was a total of 60 days. From the results, it can be seen that adding steel fibres to the concrete mix significantly reduces the expansion due to AAR. At 14 days, it can be seen that the reference mixture (0%SF) showed approximately double the expansion of samples with steel fibres. At the end of the test (i.e. after 60 days), the expansions of the different samples were reduced by the following quantities, compared to the reference mixture (0%SF) (de Carvalho et al., 2010):
- 2%SF13: 61%
- 1%SF6: 38%

- 2%SF6: 29%
- 1%SF13: 41%

Although it is clear from the results that for the SF13, an increase in steel fibre volume is more beneficial, this does not appear to be the case for SF6. An explanation for this is that apart from the fibre orientation and distribution, the content and aspect ratio of the fibres also affect the efficiency of fibres for a certain cementitious matrix. A greater aspect ratio results in a more efficient fibre (de Carvalho et al., 2010).

This experiment showed that expansion due to AAR may be significantly reduced by up to 61% by adding steel fibres to the concrete in contents of 1% or 2% (de Carvalho et al., 2010). The results can be seen in Figure 50.

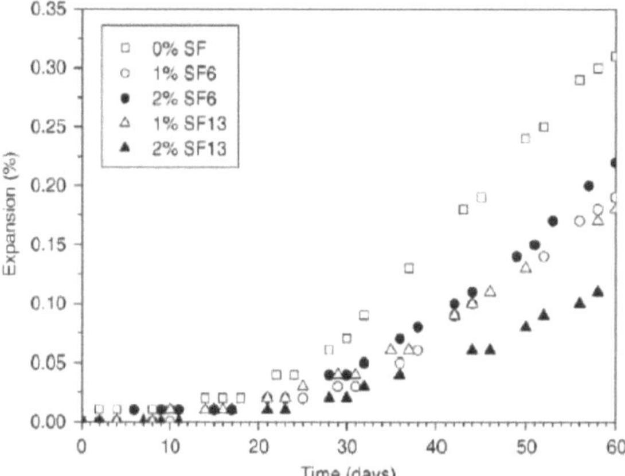

Figure 50: Expansion due to AAR (de Carvalho et al., 2010)

Appendix 6: AFNOR P 18-588 Microbar Test- Swiss Tunnels Identification of AAR

In Chapter 2, eight tunnels in Switzerland were discussed. These tunnels showed signs of AAR in the form of typical crack patterns, pop outs and excessive deposits on the concrete surface. These tunnels were assessed to verify the existence of AAR damage and to determine the extent of the damage (Leemann et al., 2005).

To assess AAR damage, visual assessments were conducted first. These assessments showed signs of typical crack patterns, pop outs and excessive deposits on the surface, which indicated AAR.

Next, 50mm diameter cores of various lengths were drilled from 16 different locations where the concrete or shotcrete showed signs of AAR in the visual assessment. During the investigation of the Swiss tunnels, four aggregates which were used to provide aggregates for seven of the coring sites were located. The potential reactivity of the aggregates from these four quarries were studied. Thin sections ($45 \times 70mm^2$) were obtained from recent productions in these quarries and the microstructure of the concrete were examined. This was achieved by impregnating them with florescent dyed epoxy resin with the use of a polarisation. Three fractured and uncoated samples of concrete were also investigated using the environmental scanning electron microscope (ESEM-FEG XL30). The ESEM had operating conditions between 15-20kV and 0.5-1.5 Torr and made use of energy dispersive X-ray spectroscopy (EDX) to identify the chemical composition of the minerals. The samples which were used to study the microstructure of the concrete were obtained from depths of 100-200mm from the surface of the concrete.

Petrographic examinations revealed that igneous rocks made up the majority of the aggregates with 53-93% by mass. These are shown in Table 31 (Leemann et al., 2005). They were deformed in various degrees and divided into granite, gneiss and schist (Leemann et al., 2005). The aggregates also consisted of major sedimentary rock types such as sandstone, limestone, siliceous limestone and slate. Minor amounts of mafic rocks were also found in the aggregates.

Table 31: Petrography of the extracted aggregate in percent by mass (Leemann et al., 2005)

Tunnel	Granite (%)	Gneiss schist (%)	Lime-stone (%)	Siliceous limestone (%)	Sands-tone (%)	Slates (%)	Mafic rocks (%)	Fragments (%)
A1/A2	41	23	-	-	6	3	5	22
B1	10	18	13	7	-	1	4	47
C1	10	22	8	5	-	1	2	51
C2	30	36	-	-	3	-	10	21
D1	21	34	6	4	5	-	1	34
F1	22	24	-	-	-	12	-	42
F2	34	24	-	-	-	14	-	28
F3	14	21	-	-	3	-	3	59
G1	16	-	18	17	18	-	3	28
G2	24	28	-	-	2	-	2	44
H1	-	44	3	-	-	6	6	41

Microscopy and microbar tests were carried out and they revealed the following (Leemann et al., 2005):

- All samples studied contained deformed quartz gains with undulatory extinction and diffused grain boundaries.
- Gneiss and schist contained varying amounts of microcrystalline quartz.
- Feldspars which were altered to sericite were found.
- Biotite, muscovite and chlorite were found in all samples.
- Six of the samples contained gel deposits in air voids and cracks. Distinct cracks patterns were observed, and they included radial cracks which ran from the aggregates into the hardened cement paste. These are shown in Table 27 (Leemann et al., 2005).
- Five other samples which didn't have gel deposits also showed the same crack pattern as above. The cracks were found in the aggregates and they had a diameter greater than 4 millimetres. The reactive aggregates were identified as schist, gneiss and granite.
- Two of the samples (F2 and D2) had cement which was extended with blastfurnace slag (about 30% of the cementitious material). The other samples had ordinary Portland cement.
- Three of the samples were investigated with ESEM and quartz with dissolution features were observed.
- Samples C2 and C3 contained feldspar and biotite with dissolution phenomena. These samples all had gel deposit in the cracks and voids. Silicon and calcium with small amounts of potassium and sodium were the main components of the gels. The gel was also present on the surface of the quartz located at the edge of the aggregate in these three samples. These are shown in Table 32 (Leemann et al., 2005).
- Nine aggregates classified as NR had an average expansion of 0.058%, and 0.040% (GI) was the lowest value. One aggregate had a classification of SR and six aggregates had a classification of R. These are summarised in Table 33 (Leemann et al., 2005).

Table 32: Indications for AAR in the microstructure (Leemann et al., 2005)

Tunnel	Reaction rims (%)	Formation of gel	Radial cracks
A1	25-50	no	no
A2	0-25	no	few
A3	50-75	yes	few
B1	0-25	no	no
C1	0-25	no	few
C2	25-50	yes	many
C3	50-75	yes	many
D1	0-25	no	no
D2	0-25	yes	few-many
E1	0-25	yes	few-many
F1	0-25	no	few
F2	0-25	no	few
F3	25-50	no	no
G1	25-50	no	few-many
G2	50-75	yes	few-many
H1	0-25	no	no

Table 33: Reactivity of the aggregate (the terms in brackets show the microbar classification determined in samples with similar petrography/samples from quarries are in italics) (Leemann et al., 2005)

Tunnel, quarry	Reactive rocks	Reactive minerals	Microbar expansion (%)	Microbar classification
A1	None	-	0.074	NR
A2	Gneiss	-	0.142	R
A3	Gneiss, schist	-	-	(NR/R: A1,A2)
B1	None	-	-	-
C1	Gneiss	-	0.070	NR
C2	Gneiss, schist	Quartz, feldspar, biotite	0.068	NR
C3	Gneiss, schist	-	-	(NR: C1, C2)
D1	None	-	0.154	R
D2	Sandstone, gneiss, schist	-	-	(R: D1)
E1	Gneiss	-	-	(R: D1)
F1	Gneiss, granite	-	0.052	NR
F2	Gneiss, granite	-	0.056	NR
F3	None	-	-	(NR: F1, F2)
G1	Gneiss, schist	-	0.040	NR
G2	Gneiss	Quartz	0.052	NR
H1	None	-	0.130	R
D1, G1	-	-	0.272	SR
F1, F2	-	-	0.065	NR
G2, F3	-	-	0.046	NR
H1	-	-	0.135	R

It was found that at a relative humidity above 80%, the expansion of the concrete due to AAR increased significantly (Leemann et al., 2005). Although the majority of tunnels usually have a relative humidity below 80%, groundwater infiltration leads to a high moisture content in concrete and shotcrete, which enables AAR to occur. In the study of the Swiss tunnels, the highest temperature reached was 21°C and therefore an assessment could not be made with regards to the effects of increasing temperatures. Laboratory tests that have been conducted have however shown that AAR is accelerated by an increase in temperature.

Assessment Method

The AFNOR P 18-588 microbar test was used to measure the potential alkali aggregate reactivity. In the case of the Swiss tunnels, this method showed a good correlation, with $R^2 = 0.94$ (for 20 samples) with the NBRI test. This test should therefore give results which are comparable to microbar tests such as the ASTM C1260 as well as the RILEM method AAR-2 (Leemann et al., 2005). The AFNOR test uses cement to aggregate (c/a) ratios of 2, 5 and 10. In Switzerland, there are no known aggregates with pessimum behaviour and therefore the value obtained from using the c/A ratio of 2 is used in the classification of the aggregates. The classification of the reactivity of the aggregates is divided into three groups: not reactive (NR), reactive (R) and strongly reactive (SR). These classification groups are in accordance with the sample expansion as follows (Leemann et al., 2005):
- NR: expansion $\leq 0.1\%$
- R: expansion $> 0.1\% < 0.2\%$
- SR: expansion ≥ 0.2

Appendix 7: Concrete Imaging – Canada
Identification of AAR

AAR is one of the most common causes of infrastructure deterioration in Eastern Canada. There has been an increasing need for improved bridge-deck condition monitoring, as a result of the rise of unprecedented concrete deterioration which has caused widespread concern about the highway infrastructure in Canada (Kabir, 2006). Concrete deterioration primarily due to AAR damage, results in structures not reaching the end of their design lives; and requiring high costs for repair and replacement. To address these concerns, concrete imaging methods are being employed to carry out bridge inspections and provide information about the current state of structures in order to assist with the determination of the remaining useful life of bridges. These concrete imaging methods are based on non-destructive testing (NDT) techniques and their use makes it possible to efficiently obtain comprehensive information regularly without disrupting traffic.

A research was carried out to analyse surface AAR map-cracking in bridge decks, using thermal, colour and greyscale imagery. The concrete deterioration features that were characterised and measured were: total amount of cracking, total length of cracks, as well as the range of crack widths. These concrete deterioration features were characterised and measured with the use of texture analysis and an artificial neural network. This research included concrete specimens which were obtained from the field, as well as concrete blocks and slabs prepared in the laboratory, all of which experienced different levels of alkali aggregate reaction induced surface cracking (Kabir, 2006).

Field Specimens

Components from various bridges with varying levels of AAR damage were used in this study. *Figure 51* shows the AAR damaged bridges in the St Lambert Lock Bridge in Montreal and a train bridge located in Sherbrooke, where components were obtained from. AAR results in swelling and eventual cracking in concrete, and the level of swelling gives an indication of the concrete deterioration, loss of rigidity and decreased mechanical properties. The components obtained from the St. Lambert Lock Bridge were severely affected by AAR, and exhibited various rates of concrete swelling. Components retrieved from bridges in Sherbrooke, including the train-bridge, all exhibited different levels of concrete deterioration.

Figure 51: (a) and (b) show AAR induced map cracking in the bridge component at the St. Lambert Lock in Montreal. (c) is the train-bridge beam in Sherbrooke (Kabir, 2006)

Laboratory Specimens

Two sets with various concrete mix designs were produced in laboratories to be used in the research. The first set was made up of three concrete blocks, of size 400 x 400 x 700 mm, which were cast with a coarse aggregate of reactive limestone. These blocks were left exposed to weather elements at the CANMET site in Ottawa, Canada. The second set comprised three concrete slabs of size 1000 x 1000 x 250 mm. These slabs were given the designation D1, D2 and D3. Slab D1 was cast with a non-reactive aggregate and sets D2 and D3 were cast with reactive limestone (Kabir, 2006). After batching and wrapping the slabs in damp terry cloths, they were then stored in ambient air at a temperature of 20 ±2°C at the GRAI laboratory at the University of Sherbrook in Québec, Canada. Tests of the amount of cracking were carried out on specimens at both laboratories. This was done due to the fact that the amount of cracking is closely linked to the level of expansion of the concrete as well as concrete deterioration indicators such as a decrease in concrete mechanical properties (Kabir, 2006). The velocities of compression (P) waves decrease when the amount of concrete damage increases. In these tests, the P-wave velocities were measured by employing the impact-echo method. The expansion of the specimens was measured by fixing stainless steel studs on the top surfaces as well as the sides of the specimens. Table 34 shows the concrete mixes, the P-wave velocities as well as the average expansion levels measured for the test specimens (Kabir, 2006).

Table 34: Concrete mixtures proportions, average P-wave velocities and expansion measurements of CANMET and GRAI specimens (Kabir, 2006)

CONCRETE MIXTURES	CANMET			GRAI		
	D1	D2	D3	D1	D2	D3
Density (kg/m^3)	2303	2303	2317	2223	2326	2340
Cement Content (kg/m^3)	423	423	425	210	390	390
Total Na$_2$O$_{eq}$(kg/m^3)	1.69	3.81	5.31	3.81	3.25	5.25
w/c	0.42	0.42	0.42	0.75	0.66	0.66
Test Measurements						
Average P-wave velocities (ms^{-1})	4909	4513	4402	3810	3590	3440
Average expansion (%)	0.025	0.374	0.383	0.000	0.060	0.100

Assessment Method

Image processing techniques are used to extract concrete damage information from structures and provide quantitative details which are not obtainable by performing visual inspection on these structures (Kabir, 2006). Visual inspections are still widely employed, however their success is heavily dependent on the skills and experience of person carrying out the assessment. Due to this, the results obtained solely from visual inspections may vary greatly and may thus produce bridge condition results which are not entirely reliable. Visual inspections are therefor used in combination with one or more non-destructive techniques or assessment (Kabir, 2006). Concrete imaging techniques may be used for this purpose.

NDT methods such as infrared thermography as well as colour and greyscale imagery have been used before. These techniques made it possible to acquire image data. The acquisition of image data alone is only the first step in the analysis, as it is necessary to further processing this data so that the information in the image data can be extracted and converted into a format which allows for easy analysis. Since concrete is a heterogeneous material, the interpretation of image data has always been challenging. It is therefore necessary to further refine imaging methods to enable the extraction of information from NDT techniques (Kabir, 2006).

A study which was carried out proposed the use of the grey level co-occurrence matrix texture analysis approach for the above-mentioned purpose. In this proposed technique, surface deterioration features in the concrete imagery are extracted, after which an artificial neural network (ANN) is used to obtain information about the concrete deterioration (Kabir, 2006). Three different image sources were used, namely: infrared thermography, visual colour and greyscale imagery. They were used on concrete specimens that experienced different levels of AAR damage. This evaluation also helped to assess how effective each imagery type is assessing the deterioration of concrete (Kabir, 2006). The three types of imagery were taken for all the laboratory specimens. For the samples taken from the field, such as the St. Lambert Lock Bridge, a digital camera was used to obtain greyscale and colour images. The thermographic images were obtained using an infrared camera.

Laboratory Samples

To perform the damage analysis, the three types of imagery of the field samples, the samples at the CANMET site as well as the samples at the GRAI site were processed using two methods, namely the grey-level co-occurrence matrix (GLCM) texture analysis method and the artificial neural network (ANN) technique (Kabir, 2006). The software to process the images were MATLAB by MathWorks Inc. and the Environment for Visualizing Images (ENVI) image processing system by Research Systems Inc. the GLCM texture analysis method was used as the first step in characterising the concrete surface damage from the images, after which the ANN technique was employed to classify and quantify the different classes of texture (Kabir, 2006). For the ANN technique used in this study, there were four input features which were: the original input image and three selected second-order features namely the mean, homogeneity and dissimilarity. Three output nodes were used for three object classes, namely wide crack, narrow crack and no crack.

The details of the processing methods are outside the scope of this book, however results will be discussed to illustrate the outcomes of using these methods. The results can be presented in two different ways (Kabir, 2006):

The results may be shown as classified images which show the spatial distribution of the different classes. In each class, every pixel is represented by a symbol or colour relating it to a specific class, as shown in *Figure 52*, for the D3 sample from the CANMET laboratory site.

 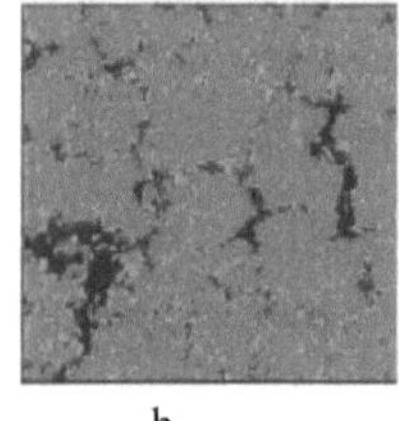 

| a | b | c |

Figure 52:Classified images of CANMET-D3 (a) greyscale, (b) colour and (c) thermographic (Kabir, 2006)

The results may be shown in tabular form, where the numbers of pixels in the whole image per class are summarised. Table 35 shows such a summary for the thermographic classifications for the CANMET and GRAI samples. The accuracy of the results obtained from the classification was then assessed, using the Kappa coefficient, as it was the index found to provide the best classification accuracy.

Table 35: Tabular representation of thermographic classifications (Kabir, 2006)

Samples	Classes	CANMET		GRAI	
		Image Pixels	**Image (%)**	**Image Pixels**	**Image (%)**
D3	Wide Crack	37356	14.25	23907	9.12
	Narrow Crack	51852	19.78	36674	13.99
	No Crack	172936	65.97	201563	76.89
	Total Pixels	262144	100.00	262144	100.00
D2	Wide Crack	9489	3.62	14078	5.37
	Narrow Crack	21365	8.15	22937	8.75
	No Crack	231290	88.23	225129	85.88
	Total Pixels	262144	100.00	262144	100.00
D1	Wide Crack	0	0.00	0	0.00
	Narrow Crack	10119	3.86	5943	1.14
	No Crack	252025	96.14	256201	98.86
	Total Pixels	262144	100.00	262144	100.00

Classifications were done for all the CANMET and GRAI samples, and the results of these are shown in Table 36. The classification accuracies for each class as well as the Kappa coefficient and the overall accuracies are also shown in Table 36.

Table 36: Classification accuracies of the CANMET and GRAI specimens (Kabir, 2006)

CANMET Samples	Thermographic			Visual Colour			Greyscale		
	D1	**D2**	**D3**	**D1**	**D2**	**D3**	**D1**	**D2**	**D3**
Kappa Coefficient	0.73	0.73	0.73	0.69	0.71	0.74	0.74	0.73	0.76
Overall Accuracy (%)	74.5	73.1	76.3	71.4	75.2	74.1	72.3	76.5	75.4
Classes	**Accuracy (%)**								
Wide Crack	81.3	79.9	83.5	78.6	78.1	78.4	76.8	78.1	78.4
Narrow Crack	79.7	78.6	81.5	77.7	70.9	77.5	76.3	70.9	77.5
No Crack	82.4	76.7	80.6	75.3	74.6	74.2	73.6	71.4	74.2
GRAI Samples	**Accuracy (%)**								
Kappa Coefficient	0.75	0.74	0.76	0.72	0.74	0.74	0.69	0.72	0.74
Overall Accuracy (%)	75.6	76.9	74.2	70.9	71.6	72.0	68.7	74.1	75.3
Classes	**Accuracy (%)**								
Wide Crack	76.7	78.1	80.0	73.4	76.1	79.9	70.4	77.1	79.7
Narrow Crack	74.7	75.2	73.4	70.1	72.9	73.4	73.7	72.9	73.4
No Crack	76.6	74.9	79.1	72.6	76.4	78.2	73.9	71.4	77.0

From the results, the following can be seen (Kabir, 2006):
- Greyscale imagery performed fairly well for the CANMET blocks, with an overall classification accuracy range of 72.3% - 76.5%
- Greyscale imagery performed fairly well for the GRAI blocks, with an overall classification accuracy range of 68.7% - 75.3%

- Visual colour imagery performed slightly worse than Greyscale imagery for the CANMET blocks, with an overall classification accuracy range of 71.4% - 75.2%
- Visual colour imagery performed slightly better than Greyscale imagery for the GRAI blocks, with an overall classification accuracy range of 70.9% - 72.0%
- Thermographic imagery produced the highest overall classification accuracies for the CANMET blocks, with an overall classification accuracy range of 74.5% - 76.3%
- Thermographic imagery produced the highest overall classification accuracies for the GRAI blocks, with an overall classification accuracy range of 75.6% - 76.9%

The classification results from all the laboratory specimen show that infrared thermography performed better than both the visual colour and greyscale imagery, and therefore the therefore the results of infrared thermography were used to quantify the different levels of AAR damage of the different specimens. Table 34 shows the results from the infrared thermography for the CANMET and the GRAI samples. The narrow and wide cracks from Table 36, together with the expansion levels from Table 34, for the CANMET and GRAI samples are shown in *Figure 53*. The average expansion levels were multiplied by of 100 for ease of comparison.

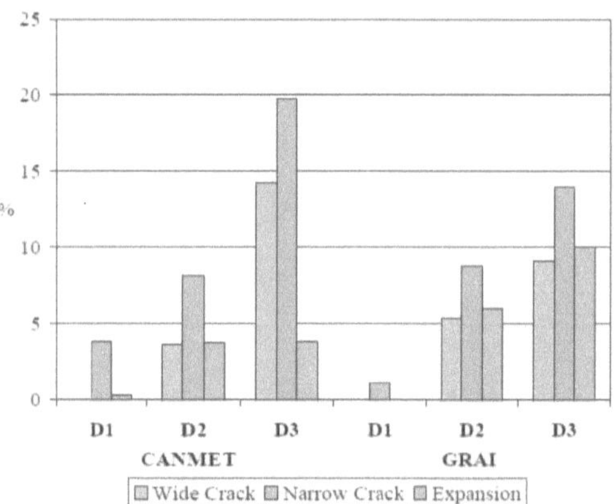

Figure 53: Comparison of concrete crack damage and expansion levels among the CANMET specimens and the GRAI specimens (Kabir, 2006)

For further analysis of the concrete surface damage, the classified thermographic images were converted into binary images, whereby the image is simplifies by assigning pixels which represent the damage in the concrete a value of 1 (black) and by assigning the background pixels a value of 0 (white) (Kabir, 2006). Manual or automated methods were then employed to add the pixels in order to calculate the total length of the wide crack, and also the average width of the wide crack. The total length of wide cracks was measured by adding all the pixels along the length a specific branch of the cracks, and this total was multiplied by the pixels resolution, which was 0.26 mm. The results in Table 37 were obtained for wide cracks (Kabir, 2006).

These results were consistent with the data shown in *Figure 53*, for the CANMET and GRAI specimens. From Table 37, CANMET specimen D1 exhibited the smallest expansion due to the fact that the concrete mix had a low alkali content. CANMET specimen D3 exhibited the biggest expansion due to the fact that the concrete mix had the highest alkali content. CANMET specimen D3 also exhibited the highest values of total length of wide cracks as well as average crack length, which corresponded with it having the lowest P-wave velocities, indicating that it had the highest level of deterioration (Kabir, 2006).

Table 37: Crack lengths and average crack widths (Kabir, 2006)

CRACK PROPERTIES	CANMET			GRAI		
	D1	**D2**	**D3**	**D1**	**D2**	**D3**
Total Crack Length (mm)	0	97.6	237.4	0	38.6	107.3
Average Crack Width (mm)	0	0.8	1.6	0	0.3	0.8

GRAI specimen D1 exhibited no wide cracks, and this corresponded with the fact that it had the lowest expansion level and this very little damage. GRAI specimen D2 exhibited a higher expansion level, while GRAI specimen D3 exhibited the highest expansion level from the GRAI specimens. *Figure 53* also shows that there is a strong correlation between the amount of wide crack damage and in the GRAI concrete specimens and the average level of expansion (Kabir, 2006).

Field Samples

Classification results were obtained as follows (Kabir, 2006):

- Thermographic, colour, and greyscale images were taken from bridge components of St. Lambert Lock
- Colour and greyscale imagery were taken of Jacques-Cartier, the Joffre and Terrill bridges and the train bridge.

  The results obtained were similar to those obtained from the CANMET and GRAI classification. As with the laboratory specimens, infrared thermography also exhibited higher accuracies than greyscale and visual colour imagery for the bridge component of the St. Lambert Lock. For the other field samples, colour imagery performed better than greyscale imagery. As with the laboratory specimens, the wide crack lengths and the crack widths were calculated (Kabir, 2006). From all the field samples, the Joffre Bridge component exhibited a wide range of crack widths, ranging from 0.15 mm to 0.3 mm (Kabir, 2006). The classified colour image of the component from Joffre Bridge is shown in *Figure 54*. Zoomed areas indicate the different ranges of cracks widths. The crack width ranges are as follows (Kabir, 2006):
- C1: 0.1mm – 0.15mm
- C2: 0.15mm – 0.20mm
- C3: 0.20mm – 0.30mm
- C4: above 0.3mm

It can be concluded that making use of the GLCM texture method and the ANN classification technique are effective in analysing surface damage due to AAR deterioration in concrete. The application of these approaches makes it possible to not only detect surface deterioration in concrete such as cracks, but also the quantification of detected defects using thermographic, visual colour and greyscale (Kabir, 2006). The classification accuracies produced by thermographic imagery were the highest for all the samples where this imagery was employed, and it therefore produced the best results. Colour imagery also produced acceptable results and it was employed in analysing damage in some field samples (Kabir, 2006). Another conclusion that may be drawn was that there was a good correlation between the amount of wide cracks in the concrete sample and its expansion level.

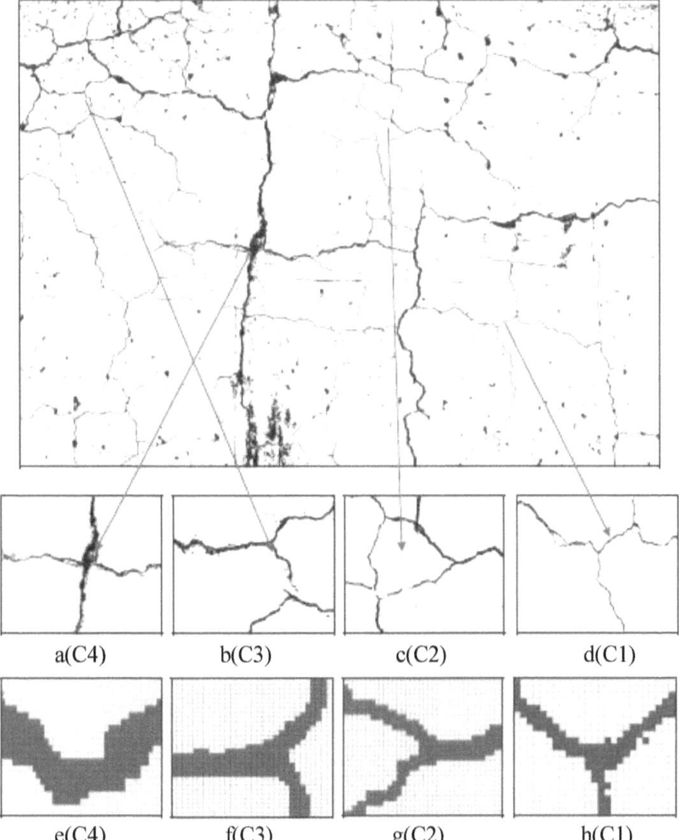

Figure 54: Different ranges of crack widths from binary image of Joffre Bridge. (a)-(d) zoomed to 1x and (e)-(h) zoomed of 20 times with a grid of 1 pixel per square (Kabir, 2006)

These methods used are less costly and less time-consuming than the conventional visual inspection methods, and therefore allow for evaluations to be carried out more frequently to supplement the conventional visual assessments (Kabir, 2006). Another advantage of employing these methods is that they allow for quantitative evaluations, such as assessing the total amount of surface damage in the images. This can then improve the quality of the concrete condition information which is obtained from conventional inspections that are performed with the aim of making decisions regarding maintenance and repairs of concrete structures (Kabir, 2006).